TRENDS IN MACROMOLECULAR SCIENCE

A

MIDLAND MACROMOLECULAR MONOGRAPHS

A series of monographs based on special symposia held at the Midland Macromolecular Institute.

Editor: Hans-Georg Elias

Volume 1 TRENDS IN MACROMOLECULAR SCIENCE

Edited by Hans-Georg Elias

TRENDS IN MACROMOLECULAR SCIENCE

Edited by
HANS-GEORG ELIAS
Midland Macromolecular Institute

GORDON AND BREACH SCIENCE PUBLISHERS
LONDON NEW YORK PARIS

Gordon and Breach Science Publishers Ltd.
41/42 William IV Street
London W.C.2

Editorial office for the United States of America

Gordon and Breach Science Publishers Inc.
One Park Avenue
New York, N.Y. 10016

Editorial office for France

Gordon & Breach
7–9 rue Emile Dubois
Paris 75014

The scientific articles in this book are also published in *The International Journal of Polymeric Materials*, Volume 2, Number 4 (1973).

Library of Congress catalog card number 73–86253. ISBN 0 677 15860 2. Printed in Great Britain.

Preface

Midland Macromolecular Institute will sponsor a series of special symposia in the field of synthetic and biological macromolecules—their chemistry, physics and technology. A typical symposium will center around one subject, an example for this will be our 1973 Symposium on "Order in Polymer Solutions", covering the conformation of macromolecules, their interaction with solvents, and their self-association. Invited speakers will deliver review lectures. We will furthermore accept a limited number of contributed short papers. The number of participants will be limited in order to encourage lively discussions. We trust that our facilities will stimulate the exchange of ideas among our guests and we hope that all will enjoy their stay in Midland.

This monograph, however, is clearly different from the ones which will follow. It does not deal with a special topic, it rather tries to present a survey of trends in Macromolecular Science. The lectures were given at the occasion of the dedication of Midland Macromolecular Institute which took place on September 28 and 29, 1972 in Midland, Michigan. I would like to take again the opportunity to express my sincerest thanks to all our distinguished speakers and the two chairmen, Professor Herman F. Mark and Dr. Raymond F. Boyer.

The Dedication Ceremonies and the Scientific Symposium were attended by over 400 scientists from 22 states, the District of Columbia, Canada, England, Germany, and Japan. They represented 51 universities and colleges and 27 companies. In view of this widespread interest in this brand new institute, we thought it appropriate to add a small chapter on Midland Macromolecular Institute. The plans leading to its foundation, its present structure and mode of operation as well as the hopes of the scientific community are well expressed in the dedication addresses.

No better can the hopes and views of the macromolecular community be expressed than in the programmatic feature lecture, given by Professor Paul J. Flory at the Dedication Ceremony. I can only hope that it is read not only by scientists but by those members of policy-making bodies to whom it was addressed in the first place: public office holders, administrators, foundation members and all those whose decisions will affect the daily life of us scientists.

Midland, Mich. HANS-GEORG ELIAS

Contents

Dedication Addresses

HERBERT D. DOAN

President, Michigan Foundation for Advanced Research

I would like to introduce our program to you by telling you briefly something of the ideas and philosophy that brought the Midland Macromolecular Institute into existence.

In fact, these ideas started to develop more than ten years ago. Their formulation into the present Institute involved so many people that I am afraid it will be necessary to stick with the ideas without referring to the many, many people who developed and influenced those ideas.

In the very early 60's it became apparent there was a migration of highly trained scientists from the Midwest to the East and West Coasts—perhaps even a greater movement than the contemporary and more widely publicized "brain drain" from England to the United States and Canada.

Now this trend—like every other trend—continued the seeds of its own destruction; but it, nevertheless, seemed that it might be useful to promote its reversal, even in a small way.

If this sounds like regional parochialism, then another factor in our thought processes was even more so. Midland has a very large number of highly trained people, many of them technically oriented. It seemed that anything we could do to develop cultural, educational and research activities would, in the long run, promote the kind of change and growth that would ensure an exciting and excellent community.

Social comment and action these days is almost entirely aimed at trying to patch up existing and severe problems—and rightly so. But it has seemed to some of us that there is also a very real need to continue and extend the excellence of communities such as Midland represents.

By doing this through science, as you people in this room understand, we have the possibility of both contributing to excellence and to the fundamental understanding and solution of known problems in ways we cannot now anticipate.

And so, in 1964 the Michigan Foundation for Advanced Research was set up supported by three Midland foundations.

Several years were spent developing, examining and even testing out ways to accomplish our basic objectives.

We surveyed the merit of one idea with scientists across the Country. You might be amused at the geographical results of this study. The West Coast people tended to say "great—try it". The Midwest scientists thought it might work, although they had their doubts about Midland as a site. The East Coast folks said flatly, "It won't work". It occurred to us these regional groups might be saying more about themselves than about our project!

In any event, in the rather long process of these examinations we were encouraged to think we were on a useful track when Government funds for fundamental research were substantially reduced. And we were even more encouraged with the potential for our project when popular opinion turned its back on technology.

There is no guarantee you are on the side of virtue by going against the tide, but there is a satisfaction in holding your finger in what you see as a worthwhile dike.

We got a lot of ideas and help and encouragement from the scientific community. Perhaps the most universal negative was the thought that we could not succeed without the immediate proximity of a good university. We have two thoughts on this: One, we see no fundamental reason why our approach will not work, particularly if we develop good cooperation with several universities; and two—and again in the long run—we see the possibility of our local educational institutions developing into one or two good universities. We will create a button and then all we have to do is weave a vest.

The idea that seemed to hold the most merit, as you know, was a polymer research institute. This plan was announced in 1967. We were able to enlist Dr. Elias as the Director, and in 1970 we started the physical construction of the Midland Macromolecular Institute.

We believe Hans has selected a staff that gives us the chance of being a significant contributor to mankind through good science.

Self-interest is often a good guide to social contribution. We believe our parochial efforts will be of benefit to Midland and the Midwest—but also to the United States and to the world.

But in isolation we are nothing. Many of you here helped in launching this Institute. We appreciate your coming here today, and we do hope being here will help inspire your continued and much needed cooperation.

HANS-GEORG ELIAS
Director, Midland Macromolecular Institute

Ladies and Gentlemen,

It is an honor and a pleasure to be the first director of Midland Macromolecular Institute. An honor and a pleasure because there are very few opportunities for very few scientists in their lifetimes like this one: few if any restrictions with respect to the type of work and a beautiful if not to say luxurious building. I know I speak as well for all the Midland Macromolecular Institute staff when I express my sincerest gratitude to the Michigan Foundation for Advanced Research for having made this opportunity available to us.

Many individuals and companies have helped us by donations, too many to be mentioned individually. I am pleased, however, to acknowledge the contribution of journals by the German National Science Foundation, represented here today by the German Consul General, Dr. Werner Montag.

It is exciting to start a brand new institute from scratch. Of course, as a scientist, I am not always excited about doing administrative business ranging from pension plans and insurance problems to snow removal and lawn care. But I am very happy to acknowledge the invaluable help I have received from the staff of the Michigan Foundation for Advanced Research, in particular from Dr. David H. Morgan, Mr. Frank Harlow, Mr. Jim Case, Mr. Jack Pyle and Mr. Frank Towsley. I wish to thank them all.

Starting a new institute means thinking about its rôle on a local, national and international level. Where shall it go? The frame was set by the Foundation: basic research, initially with a small number of permanent scientific staff and a rotating staff of postdoctoral fellows. This frame gave me, of course, great freedom in setting-up organizational plans, selecting key people and choosing activity areas. On this occasion, I want to mention the help of numerous individuals all over the country and the world who have helped me by recommending candidates and pointing out fruitful areas of research. My sincerest thanks go to a small local group of scientists who have not only given me their advice in selecting people but in numerous small details. Thank you for your help and friendship, Drs. Boyer, Alfrey and Rieke.

Scientific research means not only search for facts and a re-search of previous findings and ideas for a better understanding, it means a continuing education of the researchers themselves by the research process. This is especially true in such an interdisciplinary field as macromolecular science: ranging from synthetic organic and inorganic polymers to biopolymers, and from chemistry and physics to engineering. I was therefore looking for generalists with broad interests in the whole macromolecular field. On the other hand, our research has to be based on powerful instruments which cannot be run as black-box operations. Our senior staff should thus not only be generalists, but also experts in certain key instruments. I am very happy that Dr. Meier (formerly at Shell Development), Dr. Miller (formerly at Monsanto), and Dr. Solc (formerly at Dartmouth College) have accepted positions at MMI. Combined our research interests will cover initially such diverse areas as polymerizable detergents, poly(α-amino acids), properties of block copolymers and crystalline polymers, theory of random coils and sickle cell aenemia.

We all plan to have good cooperation not only between the groups at MMI but also with other universities and colleges and with industry. The senior staff of MMI can do (and does do) consulting to several companies. MMI is ready to do contract research work for industry and service work for universities and colleges. We plan to make available our specialized equipment for service work, especially our 260 MHz NMR instrument. We hope furthermore that scientists from universities and industry will find this place attractive for a sabbatical leave or a short stay.

Education is the second big field of MMI's activities. Since March 1972, we have had an educational program on the postdoctoral level in which we discuss fundamental problems and findings of macromolecular science in a systematic manner, using a comprehensive textbook as a frame of reference. This discussion group is limited in number, but open to our colleagues from industry. Next winter, we will introduce a polymer course on the undergraduate level. In winter 1973, we will add courses on the graduate and postgraduate level. Several nearby colleges and universities have indicated their interest in giving credit for such courses. Finally, I hope that we will have M.S. and Ph.D. candidates working at MMI which can be arranged via individual agreements or adjunct professorships.

Lecturing at MMI will not only be done by MMI staff. During the first 9 months of this year, we have already had 12 guest lectures by renowned scientists, including speakers from such countries as England, France, Japan and the Netherlands. I am very happy to have this opportunity to announce to you the creation of the first visiting professorship at MMI which has been set up by the joint efforts of the Midland section of the American Chemical Society, Central Michigan University, Saginaw Valley College, The Dow Chemical Company, Dow Corning Corporation and Midland Macromolecular Institute. The visiting professor is to give a three-week course at MMI on a subject of mutual interest to all participating organizations. I hope that we can establish a second visiting professorship soon.

International symposia on selected topics and with a restricted number of participants will be organized by MMI, as well. The first one is scheduled to take place in Fall 1973 and will cover "Order in Polymer Solutions". It will be followed by a second conference on "Polymerization in Ordered Systems" in Summer 1974. We will organize such symposia once or twice a year. The proceedings will be published in a Midland Macromolecular Monograph series.

I hope that the activities outlined will serve the local, national and international scientific community. I take it as a good indication for our objectives that so many famous scientists have agreed to deliver lectures at our scientific meeting tomorrow and that so many people have wanted to attend the Dedication Ceremonies. Thank you all for coming to MMI and to Midland.

MELVIN CALVIN

Past-President, American Chemical Society

Dr. Elias, Members of the Staff of the Midland Macromolecular Institute, and Guests:

On behalf of Dr. Max Tishler, president of the American Chemical Society and in its name as its immediate past-president, I bring you greetings and congratulations on the formal opening of the Institute today. I take a special personal pleasure in having this opportunity today to participate in what I know is only the beginning of the fruition of the long-standing dreams of many of my friends, particularly President Ted Doan and his colleagues of the Michigan Foundation for Advanced Research, Dr. Ray Boyer, their principal scientific adviser and Dr. David Morgan, their executive director. The American Chemical Society, and in fact the whole scientific community of the nation and the world, views this opening of a new private institute for basic science with great hope and expectation. This is not only for the obvious practical reason with which the ACS, and the whole scientific community is presently concerned—namely the increased opportunity for its members to do the kind of work they are best fitted to do—but for the general improvement of human welfare which this sort of activity always can have and has as its consequence.

In recent years many of us the world over have been induced to think about the place of "Science" in "Society" or even in "Civilization" itself. During the time of the renaissance and the industrial revolution, science was essentially an activity of amateurs—an intellectual activity much like the other creative intellectual activities of man. In the late 19th and early 20th centuries, it became institutionalized in academia and industry both as a philosophical resource and a source of technology for industry. However, in the last decade or so, its function in both spheres has been seriously questioned. This has been manifested in a variety of ways. Most concrete has been the skepticism with which both society and industry have considered the contributions of science to their respective needs. In a very real sense these have been of an incompatible nature—the public image being that science has been doing it a disservice in the form of activities which result in the degradation of the quality of life (pollution, crowding, etc.) and industry complaining that science is *not* providing the technology which it needs to progress.

There is a small element of truth in both of these mutually contradictory views. This has resulted in a re-evaluation, by the researchers themselves, of their proper place in both society and industry with what I believe are salutary effects in both areas. As a result of the reduction of the amount of basic scientific research now possible, the quality of the fundamental research being done, both publicly and privately, has improved and the proposals for further work have also improved. This process has been carried as far as it can go. Anything further could result in a dangerous decrease not only of quantity but of quality in an uncontrollable downward spiral.

It is especially gratifying to see the initiation of a privately financed institute for fundamental research such as this one. It seems to me indicative of the basic confidence which private resources have in the importance of basic research to the quality of life and the development of technology. If science and its resultant technology has had a part in the development of the great problems of the day—population, power, pollution and the like—it has been during the pursuit of the highest human ends, *knowledge* and its application to disease, hunger, the amenities of life—housing, transportation and communications. The solutions to these great problems of the society of men cannot reside in a "return to nature" except of the most brutal sort—with disease, starvation and death from various sources. It lies in *more* knowledge of the kind that is and will be generated in institutions such as this one—and its proper application by the industry of the world with the guidance of society.

I am confident that you will fulfill your avowed mission and help in the further development of man and his world.

CHARLES G. OVERBERGER

Macromolecular Division, International Union of Pure and Applied Chemistry

Dr. and Mrs. Elias, honored guests, devotees of macromolecular science and technology, and possible converts, I am representing here the Macromolecular Division of the International Union of Pure and Applied Chemistry. I think, on behalf of this Division, its current President, Professor Henri Benoit, its Secretary, Professor G. Smets, and myself, we certainly want to wish the new Institute every success in its endeavors in macromolecular science and technology. I could be a bit parochial also, and say that I am extremely pleased that this area has another institute in this field. I can assure the staff of MMI, if I may put on another hat for a moment, that they will receive in every way all the cooperation from the Macromolecular Research Center of the University of Michigan and the Office of the Vice-President for Research of the University of Michigan.

I think also that I would add my congratulations to the Michigan Foundation for Advanced Research and the key people who were involved in the development of this concept. It was not easy and required patience; and Ted Doan as Chairman, Ray Boyer as Scientific Adviser, and many others did a most notable task here in organizing and obtaining the necessary backing, enthusiasm and in finally obtaining this very excellent director that we have for this new institute.

I do want to say something very briefly about the Macromolecular Division of IUPAC. One thing about being a speaker near the end of the group is that you don't have to speak so long because all of the other people have said the important things. But there is an aspect to this which I think is worth reflecting on. IUPAC is a diffuse organization. There are six divisions in it: organic, physical, inorganic, analytical, applied, and macromolecular. Now the reason that's so is because the people in this audience, in this group, are basically very much research-oriented people who have an interest in an outlet for themselves in an international organization. When the macromolecular group was a commission of the Physical Division, as Professor Mark will remember well, it was probably one of the most active commissions of all time and there is no question about the fact that this interest and international enthusiasm has been a major factor in why IUPAC has a Macromolecular Division. It is a truly interdisciplinary division that crosses a lot of borders; it is an excellent interface between industry and academia. So I would ask you all to remember this and remember our constant enthusiasm and interest in macromolecular science or any peripheral areas. We hope this institute will be another outlet for this enthusiasm. And I am sure, Hans, that you have the confidence of everyone here and best wishes in the audience as well.

Finally, I would simply say that I think the advent of macromolecular science in the universities and in institutes of this kind, clearly indicates not only the educational needs for this type of training but also the continued interest that the industrial community will give to this growing area of science.

Thank you.

WERNER MONTAG

Consul General, Federal Republic of Germany

President Doan, Professor Elias, distinguished hosts and guests:

After arriving here from my Consulate General in Detroit at 10 o'clock, I had a most instructive tour of the Dow Chemical Company with Dr. David Morgan as my Cicerone, followed by a very pleasant get-together discussing Germany with the Midland Rotarians.

Now I have the honor and pleasure to represent the Federal Republic of Germany at the Dedication Ceremonies of the Midland Macromolecular Institute.

I feel relieved that I do not have to represent here Chemistry and Physics—my scientific career in these fields having come to quite an early end at the preliminary examination—Vorexamen—in 1940 . . . what obviously qualified me to become a consul. Still I am grateful having been able to hear at the Technical University of Berlin such world renowned scientists as Professor Geiger, Vollmar and others who contributed the Geigercounter and other discoveries to the development of science.

From this you will understand my very personal involvement in felicitating, on behalf of the German Government, the Dow Foundation, Gerstacker Foundation and Strosacker Foundation, the Michigan Foundation for Advanced Research and the many personalities who made MMI come to life. I feel comfortable standing before you not with completely empty hands. The Alexander von Humboldt-Stiftung and the Deutsche Forschungsgemeinschaft have contributed to the library of this Institute, which is to attract scientific leaders from around the globe, 140 volumes of Germany's Macromolecular Chemistry with the value of 6,313 German Marks, or 2,000 Dollars. I entrust them to my German compatriot Professor Hans-Georg Elias, coming from the world-known Technical University of München—my second Alma Mater—and Swiss Federal Institute of Technology (ETH) at Zürich, Switzerland.

Let me close my remarks with a few words from the physician and Nobel Prize winner Gerhard Domagk whose chemo-therapeutic discoveries have preserved the health and life of countless individuals. He says: "What is really important in this world? That we individuals get along with each other, try to understand and help each other as best we can. For us physicians that is natural. Why shouldn't it also be possible for all other people? Don't tell me this is Utopia! Every discovery was considered utopian. Why should we first wait for another measuring of powers—we really have suffered enough to have become wise. But it is comfortable to cling to old customs; more comfortable to follow violent rulers, cholerics, paranoiacs, and other mentally disturbed individuals instead of thinking for oneself and looking for new ways of reconciliation instead of mutual destruction".

These are clear words. They come from a man who fights for life, from a man who does not merely hope to save human beings, but puts all his knowledge and ability to work for it.

Thank you for your attention.

Midland Macromolecular Institute

Midland Macromolecular Institute (MMI) is a division of the Michigan Foundation for Advanced Research (MFAR). MFAR, a private operating, non-profit foundation, was founded in 1964 by three Midland, Michigan-based private foundations: Herbert H. and Grace A. Dow Foundation, Rollin M. Gerstacker Foundation and Charles J. Strosacker Foundation.

After extensive investigation by a group of local scientists, in November of 1967, MFAR announced plans to sponsor a laboratory, located in Midland, Michigan, to be devoted to basic research and study in the field of macromolecular science. To implement these plans, MFAR then selected an Advisory Council composed of

Dr. T. Alfrey, The Dow Chemical Company, Midland,
Professor E. Baer, Head, Division of Macromolecular Science, Case Western Reserve University, Cleveland, Ohio,
Dean Richard U. Byerrum, Michigan State University, East Lansing, Mich.,
Professor Melvin Calvin, Director, Laboratory for Chemical Biodynamics, University of California at Berkeley,
Professor Paul Flory, Chairman, Department of Chemistry, Stanford University, Stanford, Calif.,
President Brage Golding, Wright State University, Dayton, Ohio,
Professor Charles Overberger, Vice-President for Research, University of Michigan, Ann Arbor, Mich.,
Professor Calvin Stevens, Chairman, Department of Chemistry, Wayne State University, Detroit, Mich.

After their consultation and many discussions with other leading macromolecular scientists, MFAR decided to set up MMI as an open national laboratory dedicated to the advanced training of scientists. It is hoped that MMI can help strengthen ties between academia and industry by either providing research facilities for industrial and university scientists on sabbatical leave or by cooperative research programs.

Midland Macromolecular Institute is a tri-level building (Figures 1 and 2). Two of the floors contain laboratories, offices and the library, whereas the third floor serves for mechanical utilities.

The floor plan is unique and very effective (Figures 3 and 4). A library forms the center of the Institute acting as mixing center for informal gatherings and

FIGURE 1 Midland Macromolecular Institute as seen from the west.

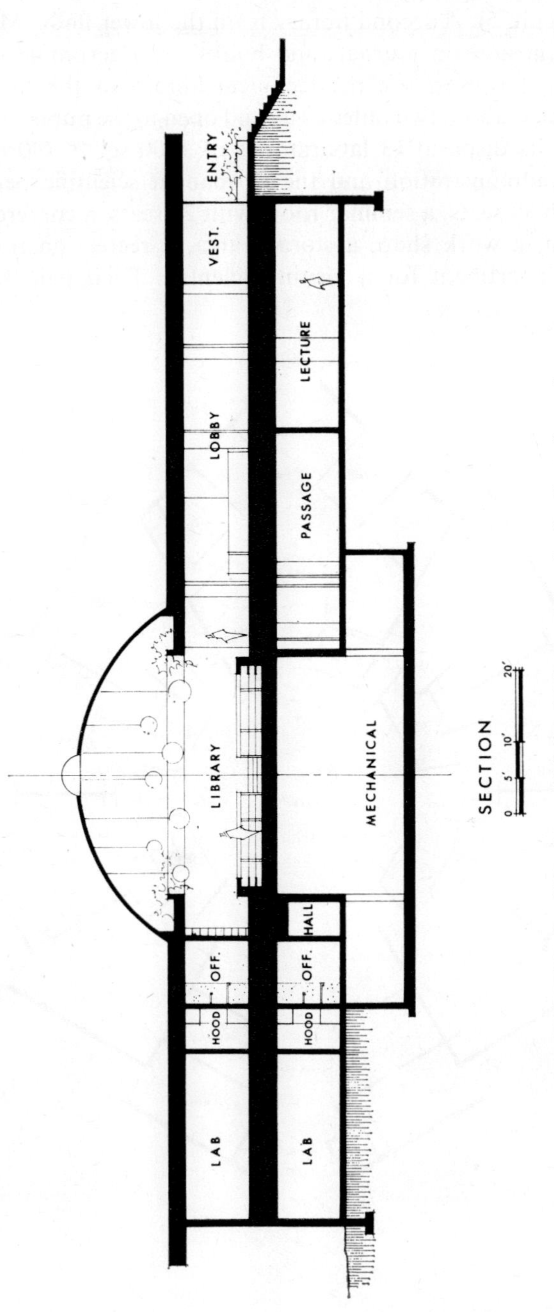

FIGURE 2 Side view of Midland Macromolecular Institute from the west.

discussions (Figure 5). A second library is on the lower floor. MMI's library naturally concentrates on journals and books on Macromolecular Science. MMI staff can, however, use the technical library of the Dow Chemical Company, which is about two miles away and open to the public free of charge.

MMI has at its disposal 13 laboratories of 1000 sq. ft. (90 m^2) each, 14 offices for the administration and the permanent scientific personnel, one lecture hall with 99 seats, a seminar room with 20 seats, a conference room, a computer room, a work shop, a storage area, a recreation room/cafeteria and a built-in apartment for a visiting scientist. Each pair of laboratory

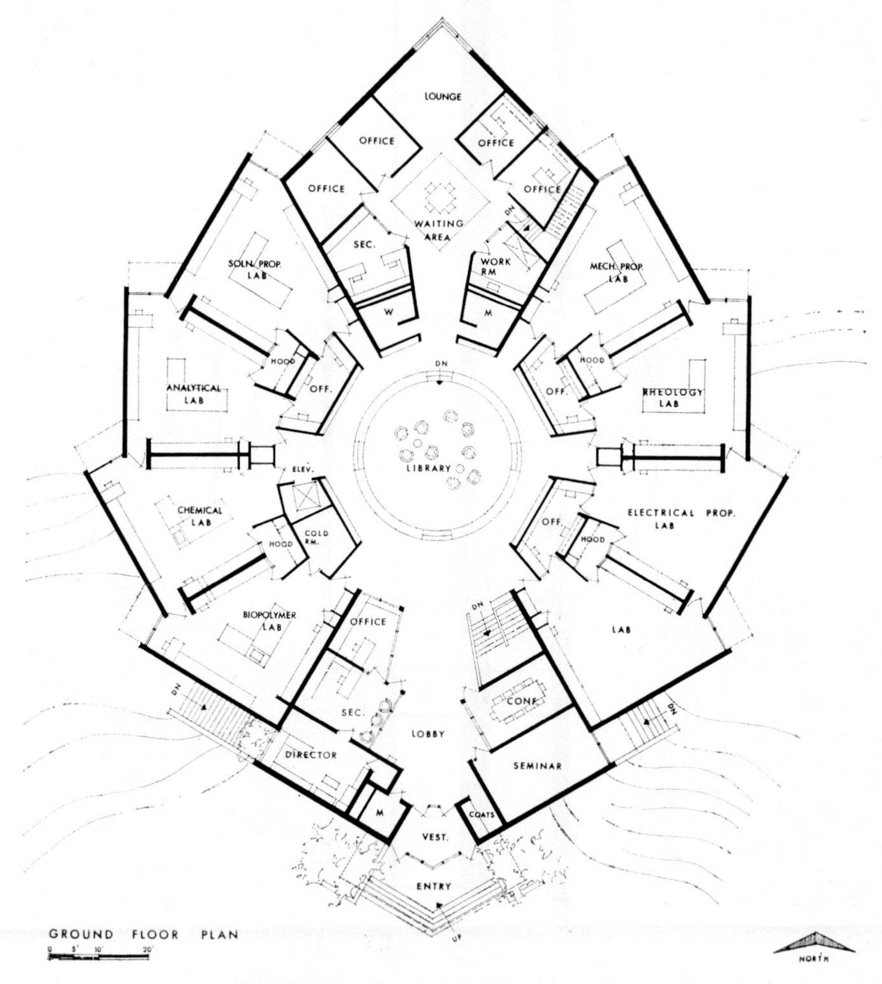

FIGURE 3 Floor plan of the upper floor.

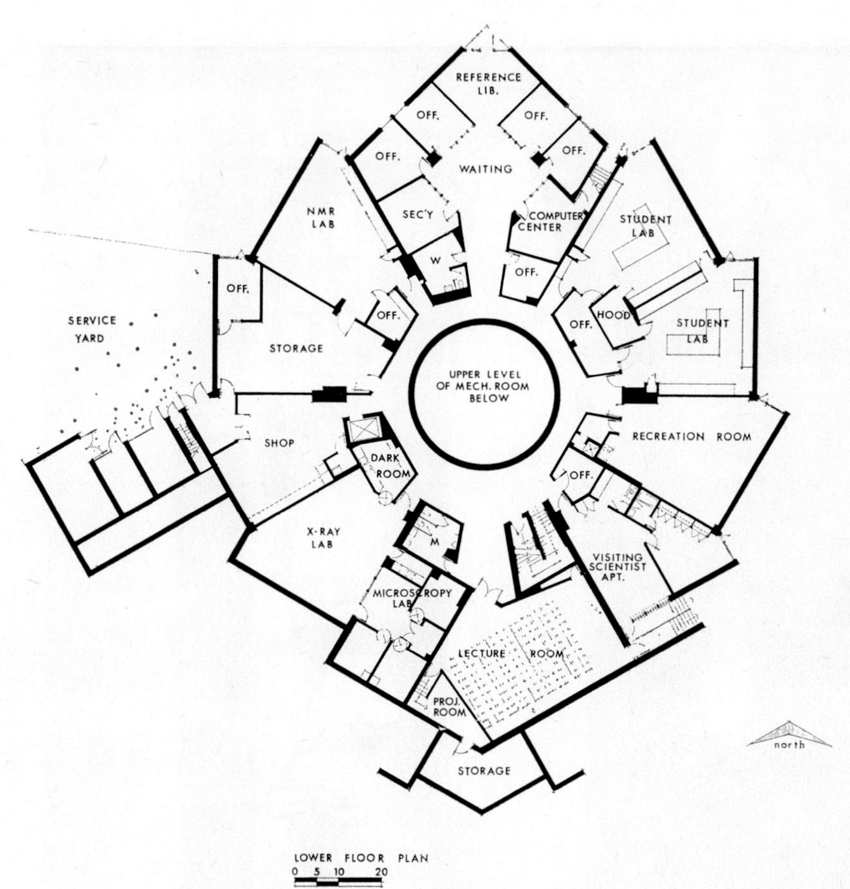

FIGURE 4 Floor plan of the lower floor.

modules shares a hood and an office for three post-doctoral workers. The built-in apartment consists of a living room, bedroom, kitchen, bathroom and office.

MMI is equipped with ultra-modern instruments: a 260 MHz nuclear magnetic resonance spectrometer with Fourier transform equipment for the measurement of various nuclei, a UV/visible spectrometer, two infrared spectrometers, a circular dichroism/optical rotatory dispersion instrument, differential scanning calorimetry and thermogravimetry, a laser Raman spectrometer, a X-ray diffractometer for broad and small angle measurements and a transmission and scanning electron microscope. Solution properties can be measured via membrane and vapor phase osmometry, light scattering, analytical ultracentrifugation, gel permeation chromatography,

FIGURE 5 Library.

and capillary and rotational viscometry. Mechanical properties can be determined with Instron, Rheovibron and High Speed Testing instruments. Polymers can be processed by calendering, injection molding, melt and solution spinning and with a Brabender.

The scientific activities are directed by a staff of five senior scientists:

Dr. Hans-Georg Elias, Director of MMI and Adjunct Professor of Chemistry at Michigan Technological University (from Swiss Federal Institute of Technology at Zürich (ETH)),

Dr. Robert J. Kostelnik, Senior Research Associate (from Carnegie-Mellon University),

Dr. Dale J. Meier, Senior Research Scientist (from Shell Development),

Dr. Robert L. Miller, Senior Research Scientist (from Monsanto Research),

Dr. Karel Šolc, Research Scientist (from Dartmouth College).

The senior scientific staff works together with the permanent technical staff and the rotating scientific staff of post-doctoral fellows, Ph.D. and M.S. candidates. It is free to select its problems which at present cover stereocontrol of polymerization, association in dilute solution, shape of random coils, properties of block copolymers, morphology of crystallizing polymers, interaction between DNA and small molecules, and sickle cell aenemia.

MMI offers a complete course program in macromolecular science. Courses cover the constitution, configuration, conformation and supermolecular structure of polymers, their synthesis and their reactions, their mechanical, rheological, electrical, optical and surface properties, their processing, and the properties of industrially and biologically important macromolecules. The courses are taught by the senior scientific staff and by visiting professors. MMI furthermore sponsors an extensive guest speaker program and conducts additional seminars on its own research. Several universities offer credit for the courses given at MMI. M.S. and Ph.D. thesis work may be carried out at MMI in cooperation with other universities. MMI itself is not a degree granting institution, however.

The Challenge to Macromolecular Science

PAUL J. FLORY†

Department of Chemistry, Stanford University, Stanford, Cal. 94305

Brief historical review of the development of macromolecular science and some of the challenges for present and future research.

The adaptation of polymeric materials for the artifacts of man originated in antiquity. Techniques for processing fibers, hides, wood, rubber and naturally occurring resins have precedents that predate recorded history. These are amongst the oldest technologies that have evolved over the centuries and up to the present. By contrast, the *science* of polymers, or macromolecules, made a late appearance—late, that is, relative to other branches of science. The modern science of molecules that we call chemistry had its beginnings around 1860. The foundations of quantum theory and quantum mechanics were established from 1900 to 1925, the theory of relativity from 1905 to about 1920. The basic concepts underlying macromolecular science were not put forward until after these epoch-making advances had determined the course and content of chemistry and physics for decades to come.

The motif of macromolecular architecture—the concatenation of atoms, or groups of atoms, to form covalently linked chains of great length—did not achieve widespread acceptance as the structural principle pervading virtually all polymeric substances until 1930 and doubts concerning its reality lingered for some years thereafter. Polymeric chemical structures had been suggested—even advocated—as much as 70 years earlier. However, for reasons difficult to grasp in retrospect, linear macromolecular structures were rejected in every instance. Instead, small cyclic structures were ascribed to substances

†Address delivered at the Dedication of the Midland Macromolecular Institute, September 28, 1972.

now recognized as polymeric, in the molecular sense of this term. Thus, rings of three to twelve bonds were proposed for proteins, rubber, cellulose, starch, and various synthetically prepared polymers. The preoccupation of chemists in general with molecules representable by concise formulae conveniently expressed on the printed page, seems to have generated an overpowering conviction that all substances should be so constituted. The much larger chemical structures suggested from time to time for the most commonplace of substances, and the most important ones in the environment of man, were consistently cast aside, probably on this account.

Not until the late 1920's was it clearly established that cellulose and starch are high polymers consisting at the molecular level of long chains now known to be the universal theme amongst all polymeric structural materials, natural and synthetic. The epic of natural rubber is similar: determination of the chemical structure of the unit at the beginning of this century, appearance of a variety of cyclic formulae, and finally, in the late 1920's, the belated acceptance of the polymeric chain formula.

The polypeptide hypothesis regarding the chemical structure of proteins put forward by Hofmeister and by Fischer about 1902 did not gain widespread acceptance before 1930. As late as the mid 1920's, leading organic chemists of the period vied with one another in matches of ingenuity to invent esoteric cyclic structures for proteins, all seemingly in a vain effort to avert adoption of the polypeptide chain structure as the pattern for life-giving proteins, the abundance of chemical and physical evidence in support of the polypeptide structure notwithstanding. Evidently the vision of a polymeric structure for materials of such supreme importance was repugnant. Whether the polypeptide chain was regarded as inherently objectionable or merely something to be opposed because it departed from the structures that were in the mainstream of chemists' endeavors is a matter for conjecture. Bear in mind that Fischer had completed his *tour de force* in synthesizing a polypeptide of 30 units during the first decade of the century. Yet, even he clung to the belief that proteins consisted of chain molecules no longer than he had synthesized, a number of these being "colloidally" aggregated to form the protein "particle".

The saga of polynucleotides is even more striking. According to opinions that went essentially unchallenged until the mid 1940's, they were tetrameric. Only in the 1950's was it established that they consist of long chains of nucleotide units numbering as many as hundreds of thousands.

Evolution of ideas regarding synthetic polymers followed a similar course. Synthetic polymer chemistry developed rapidly once their molecular nature was understood and hence the skills of the chemist could be turned systematically to their preparation in ever increasing variety. The influence of the pioneering work of Staudinger and Corothers in this period cannot be overestimated. Their work placed the macromolecular hypothesis on a firm footing and at the same time elaborated methods of measurement and synthesis.

It is abundantly evident that long chain structure as the principle common to all substances we now know to be macromolecular was accepted with great reluctance. The aversion to polymeric formulae is well documented. Reasons for it are not. The best that can be ventured is that they were begotten of a combination of esthetics that place a premium on (apparent) simplicity, and accidents of history that directed interests and efforts in other directions.

It is important to observe that macromolecules are not clearly demarcated from their analogs of lower molecular weight. In every series of macromolecular homologs, species of intermediate chain length occur. The chemical bonds joining atoms in macromolecules are not discernibly different from those in small molecules; they are described by the same geometrical parameters. Any sharply drawn distinction between the domain of the molecules that are commonplace in the laboratory and the macromolecules that are commonplace everywhere else must necessarily be arbitrary. Hence, the science of macromolecules is fundamentally coextensive with the rest of chemistry, and with some of the ramifications of physics too. Not to be overlooked, however, is the inescapable fact that definite alterations in concept and theory are required for the understanding of macromolecules. Knowledge in the domain of giant molecules is not to be gained by simple deduction from concepts and rules gathered for small molecules. Modifications in viewpoint and method are required, and in some respects a different set of guidelines must be adopted.

The relatively late emergence of macromolecular science has had an important consequence that weighs heavily on present and future. The basic pattern of chemistry as a discipline, with its traditional subdivisions, became established prior to the appreciation of the molecular nature of polymers, and prior to the appearance of theories and generalizations describing their behavior. Macromolecules were therefore left out of the syllabus in its formative years. Only a few desultory remarks on colloids found a place in the textbooks, and these were generally misleading, if not wrong altogether. The subject has not gained entry since. Irrespective of one's views on the proper place for polymer chemistry and physics in the curriculum, its virtual absence therein at present is a fact to be reckoned with. It is a reality to be taken into account in charting the course of the Midland Macromolecular Institute.

The relevance of polymer science to industrial technology is self-evident, and, I think, widely recognized. I have already alluded to the close temporal correlation between the emergence of polymer science and the proliferation of synthetic polymeric materials. If the former did not in fact furnish the main impetus for the latter, at least it has played a major role in bringing into being what posterity may refer to as the age of polymers. Incidentally, I also believe it is a valid assertion that pure and applied polymer science continue to enjoy a mutually fruitful relationship.

The relevance of polymer science—the *same* science—to biology seems not as well appreciated. Custom, and perhaps prejudice rooted in the never-dying vitalistic view of things living, seem to sustain the impression that biopolymers are unrelated to the technological members of the polymer family. To be sure, biopolymers have acquired certain attributes in the course of their evolution that are not shared by their less refined relatives. But there are basic properties and behavioral patterns shared by all polymers: the thermodynamics of their solutions, rules governing their spatial configurations, manifestation of high elasticity under suitable conditions, crystallization, etc. The tendency to relegate biopolymers and technological polymers to different categories works to the detriment of the fuller understanding of both. Much stands to be gained from the broader view that embraces both of these artificially differentiated categories. If molecular biology is to rest on a sound *molecular* basis, then logic dictates that that basis be provided by the science of polymeric molecules. For this purpose a polymer science concerned with the fundamentals of macromolecular behavior in the broadest sense is required.

There is a further role for macromolecular science, and one for which it is especially well suited. I refer to the communication between science and the public. Most subjects at the forefront of science pertain to matters abstruse and quite remote from the interests and awareness of the nonspecialist. Chemistry is illustrative: its compounds, theories, and reactions offer relatively few opportunities for cultivation of interest on the part of the public at large. Polymeric materials, being exemplified in profusion in a multitude of articles of commerce, not to mention biopolymers too intimate to require mention, are ideally suited as subjects for communication of science to a wider audience. It is here that polymer science enjoys a special opportunity by virtue of its subject and purview.

The opening of the Midland Macromolecular Institute comes at a time of rich opportunities for contributions to science, to industrial technology, and to biology and medical science. I have tried to indicate some of these opportunities without, however, specifying particular directions of scientific inquiry in need of attention. To attempt the latter would be at the hazard of giving undue weight to the interests and prejudices of one person. Suffice it therefore to convey the conviction that macromolecular science still has a long way to go if it is to meet the challenges I have indicated. Much basic and creative science needs to be carried forward. Coherence of the subject through systematization of existing knowledge is of ever increasing importance as research spreads into new directions. At the same time, the importance of cultivating connections between macromolecular science and related fields should not be overlooked.

Finally, I would stress that genuine science is not an activity that can be directed by external control. Widespread opinions to the contrary notwithstanding, it is not an activity that is responsive to the needs of society—at least as society at large envisages its needs. I agree that original science—and invention too—have been influenced in every age by the contemporary scene and mood. But creative science and invention must innovate and initiate. In the highest expressions of their capacities, they do not function as agents for providing means to ends selected by an external body, whether it be a committee, directorate, or agency of government. It was not in response to a managerial decision, an Act of Congress, a referendum, a rally, or even the FCC that the telephone was invented. The same is true of the electric light bulb, the airplane, or the discovery of penicillin. Dalton and Laviosier did not undertake their researches under orders or in response to requests placed before them.

It is therefore important that the Midland Macromolecular Institute be self-directed. It should be accorded a wide freedom of choice in the direction of its investigations. But this is only one side of the coin; the admonition has its concomitant. The senior scientific staff must accept the responsibility to conceive researches in the vanguard of macromolecular science—researches that hopefully will prove to be of the foremost significance. The extent to which they succeed in meeting the challenge of this responsibility will be the measure of success of this Institute. We wish them well!

Organic Chemistry of Macromolecules†

CHARLES G. OVERBERGER and K. N. SANNES
Department of Chemistry and the Macromolecular Research Center, The University of Michigan, Ann Arbor, Michigan 48104

Polymers as reagents in organic synthesis may offer certain advantages over low molecular weight reagents. Examples given include cyclization reactions, Dieckmann reactions, acylation and alkylation of active methylene esters, intrapolymeric reactions, hydrogenation of cyclic olefins, and hydrolysis of esters. The influence of cooperative effects and hydrophobicity on the self-acceleration and allosteric effects on some of these reactions is discussed.

1 POLYMERS AS REAGENTS IN ORGANIC SYNTHESIS

1.1 Introduction

One of the first applications of a macromolecule as a "solid support" in organic synthesis has served to revolutionize the field of peptide fabrication. Condensing specific sequences of amino acids onto a growing peptide, which is attached to an insoluble polymer, has, at least in principle, changed peptide synthesis from one of tedious, multi-faceted separations to a procedure which has been automated. No further account[1] of this field will be given in this report except for certain studies which apply to a topic under discussion.

Instead, this section of the report contains what we thought to be novel applications of polymeric reagents in synthetic organic chemistry. No attempt at inclusiveness was made.

1.2 Proximity effects

1.2.1 *Dilution or separation of reactants* The ability to control the relative rates of competitive reactions in a given system constitutes a good portion of

†Lecture at the Scientific Symposium at the occasion of the Dedication of Midland Macromolecular Institute, September 29, 1972.

successful synthetic chemistry. In particular, cyclization reactions are successful only when the rate of the intramolecular ring closure reaction dominates over the rate of the competing intermolecular coupling reaction. Traditionally, high dilution of the reactant has been used to reduce the rates of the latter reaction and, thereby, increase the amount of desired cyclic product which is formed. Obviously there are limitations involved with this method.

Another method of limiting the rate of intramolecular reactions consists, in theory, of anchoring each molecule to be cyclized to a functional group along a polymer chain at intervals that preclude intermolecular (strictly speaking it would now be "intrapolymeric") reactions.

The ring closure of amino acids attached to a polymeric support establish the feasibility of this type of reaction.[2,3] As outlined in Scheme I, the N-blocked

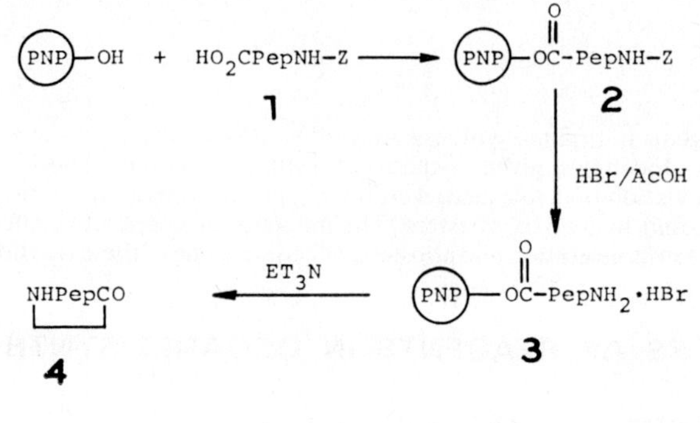

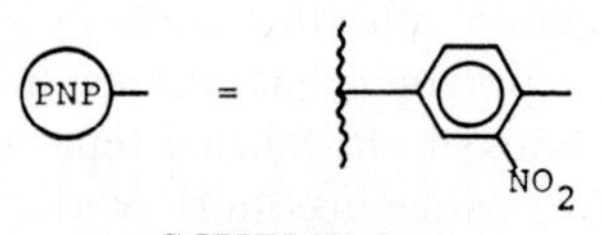

SCHEME I

amino acid, **1**, is attached to poly-3-nitro-4-hydroxystyrene to form the polymer ester **2**. Removal of the blocking group affords amine **3** which, upon treatment with triethylamine, undergoes cyclization with simultaneous separation of the newly formed cyclic peptide, **4**, from the insoluble macromolecular support.

Another study has demonstrated the feasibility of performing Dieckmann reactions on diesters which are bonded to solid supports.[4] Potassium salts of ^{14}C labeled pimelate half esters were reacted with either chloromethylated polystyrene or benzyl chloride to form compounds **5a**–**5e** (Scheme II). Treatment of **5a**, **5c** or **5e** with base afforded 15–46% yields of cyclized products.

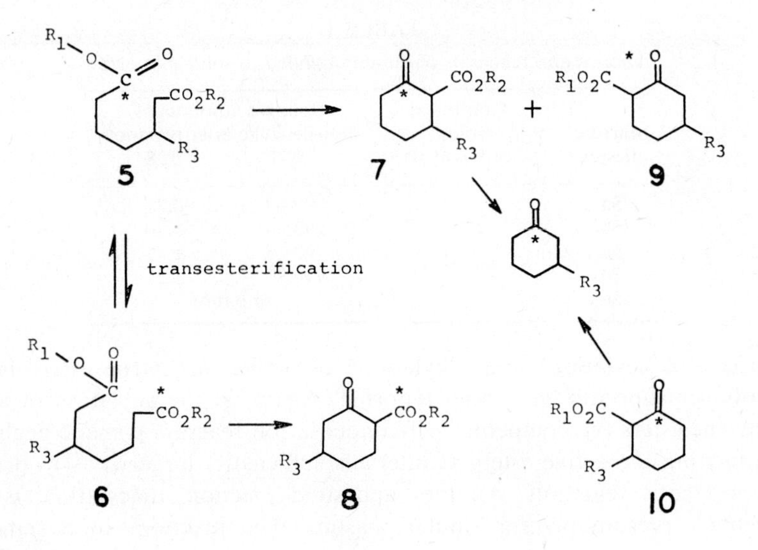

5a – 10a,	R_1 = P *p*-$C_6H_4CH_2$,	R_2 = $C(C_2H_5)_3$,	R_3 = H
5b – 10b,	R_1 = $C_6H_5CH_2$–,	R_2 = $C(C_2H_5)_3$,	R_3 = H
5c – 10c,	R_1 = P *p*-$C_6H_4CH_2$–,	R_2 = $C(C_2H_5)_3$,	R_3 = C_2H_5
5d – 10d,	R_1 = $C_6H_5CH_2$–,	R_2 = $C(C_2H_5)_3$,	R_3 = C_2H_5
5e – 10e,	R_1 = P *p*-$C_6H_4CH_2$–,	R_2 = C_2H_5,	R_3 = H

* = ^{14}C label.

SCHEME II

There are additional advantages offered with the solid support system that warrant discussion. For example, **5b** affords two cyclic products **7b** and **9b** when treated with base, which contain the alkyl and benzyl esters, respectively (Table I). Also, the group R_3 occupies a different relative position on the ring in **7b** and **9b**. When **5a**, for example, is reacted with base one might still form **9a**. However, in this case the cyclic compound **9a**, is still attached to the polymer support and is separated from **7a** during the filtration of the polymer from solution. This eliminates the need for the separation of the two similar keto esters as is the case in reactions **5b** and **5d**. Evidently a hindered ester (triethylcarbinyl in **5a** and **5c** instead of ethyl as in **5e**) must be used to prevent transesterification. Once this has been done, this procedure can serve very nicely in the preparation of specifically labeled ring compounds. As mentioned, this method can probably be extended to systems with various ring sizes.

TABLE I

Dieckmann reactions of diesters bonded to solid supports

Starting diester	Combined mole % of **9** and **10**	Relative amounts of non-benzylic ester products % **7**	% **8**
5a	—	99.69	0.31
5b	10	98.30	1.70
5c	—	97.25	2.75
5d	21	96.80	3.2
5e	—	mixture	

Successful acylation[5] and alkylation[6] of active methylene esters bound to polymer supports have been reported. Again, in the acylation of active methylene esters the competing self-condensation reaction is made negligible by attaching the active esters at intervals sufficiently large so as to prevent intrapolymeric reactions. Another undesired reaction, diacylation, is also prevented, presumably for similar reasons. The reactions in Scheme III

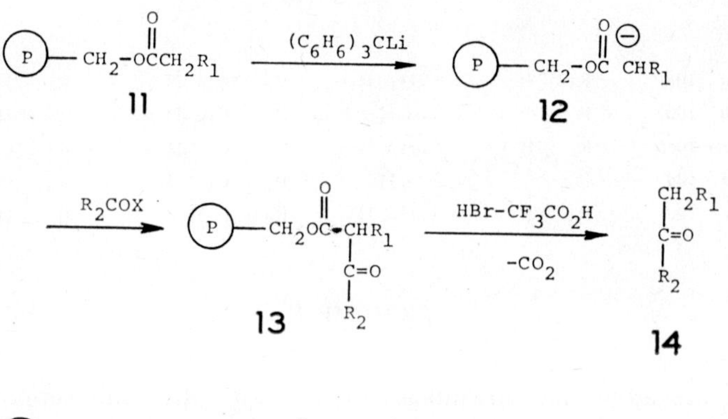

(P)— = polystyrene

R_1 = H, C_6H_6

X = Cl, OCR_2 (with C=O)

11a – 14a,	$R_1 = C_6H_5$,	$R_2 = p\text{-}NO_2C_6H_4COCl$
11b – 14b,	$R_1 = C_6H_5$,	$R_2 = p\text{-}BrC_6H_4COCl$
11c – 14c,	$R_1 = C_6H_5$,	$R_2 = (\alpha\text{-}C_{10}H_7CH_2CO)_2O$
11d – 14d,	$R_1 = H$,	$R_2 = p\text{-}NO_2C_6H_4COCl$

SCHEME III

provided a single ketone in each case (see Table II). Analogous reactions without the polymer support resulted in several ketonic compounds. Reasonably, it was observed[5] that if the concentration of ester bound to the polymer becomes too high the intrapolymeric reactions begin to interfere. The successful acylation reactions employed 0.1–0.3 mmoles of ester/gram of polymer and competing reactions were evident at concentrations of 1.5–2.0 mmol of ester/gram of polymer.

TABLE II

Ketone formation from esters bonded to solid supports

Starting material	R_2COX	Product	Yield, %	Recovered starting acids, %
11a	p-$NO_2C_6H_4COCl$	**14a**	43	40
11b	p-BrC_6H_4COCl	**14b**	40	45
11c	$(\alpha$-$C_{10}H_7CH_2CO)_2O$	**14c**	40	55
11d	p-$NO_2C_6H_4COCl$	**14d**	20	45

The alkylation of isobutyric acid[6] which was esterified onto a nonbenzylic support, as depicted in Scheme IV, afforded only benzyldimethyl acetic acid (**18**) and unreacted isobutyric acid in 20 and 80% yields, respectively.

1.2.2 *Forced combination of moieties connected to the same macromolecule* In contrast to the effect obtained by separating interreactive molecules by bonding them remotely on a macromolecule, a system could, in principle, be constructed with two types of molecules held in close proximity in order to promote this "intrapolymeric" reaction. In practice this type of proximity effect is involved in the reaction outlined in Scheme V.[7] A small percentage (2.3–25%) of chloromethylated polystyrene groups were esterified with a limited amount of an enolizable acid, **19**, and the remainder of the groups with

TABLE III

Ketone formation from enolizable esters bonded to solid support

Enolizable acid	Non-enolizable acid	Yield, %	Analogous reaction in solution
$CH_3(CH_2)_6CO_2H$	p-$ClC_6H_4CO_2H$	35	30
$C_5H_5CH_2CO_2H$	$C_6H_5CO_2H$	45	—
$C_6H_5(CH_2)_2CO_2H$	$C_6H_5CO_2H$	85	42
$CH_3(CH_2)_4CO_2H$	$C_6H_5CO_2H$	95	—
$C_6H_5(CH_2)_2CO_2H$	p-$ClC_6H_4CO_2H$	85	20

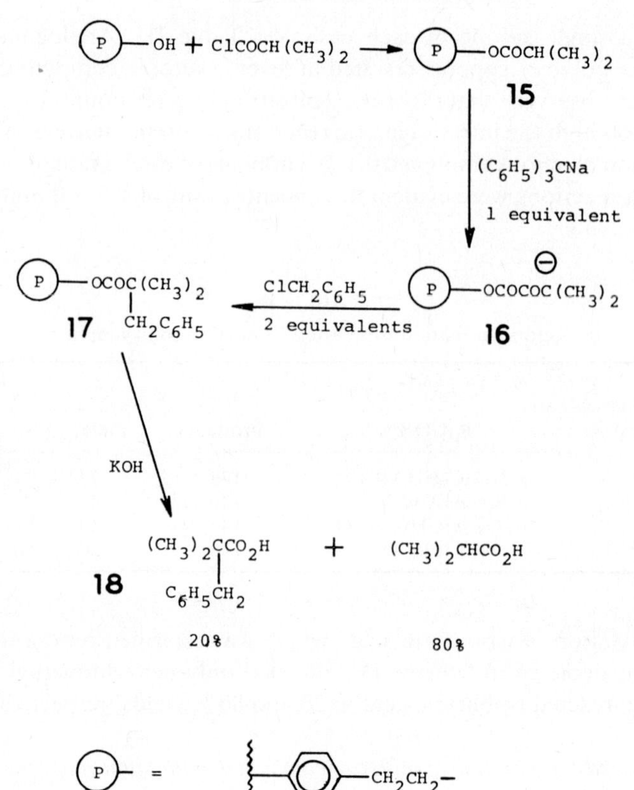

SCHEME IV

a nonenolizable acid, **21**, to give the mixed ester **22**. The addition of base results in internal condensation to form **24** which after esterolysis and decarboxylation provided *only* ketone **26** and starting acids **19** and **21**. Analogous reactions under similar conditions but without the polymer support lead to a more complex mixture of products. The intrapolymeric nature (**23** → **24** in Scheme V) of the reaction is evidenced by the fact that no ketone was observed when two batches of polymeric esters—one with an enolizable and one with a nonenolizable ester—were reacted and worked-up under conditions similar to those used for **22**. This scheme is in line with the fact that a larger ratio of **21**:**19** in **22** results in higher yields.

In the solid-phase synthesis of peptides, side-reactions limit the length of the peptide chain that can be obtained in purified form.

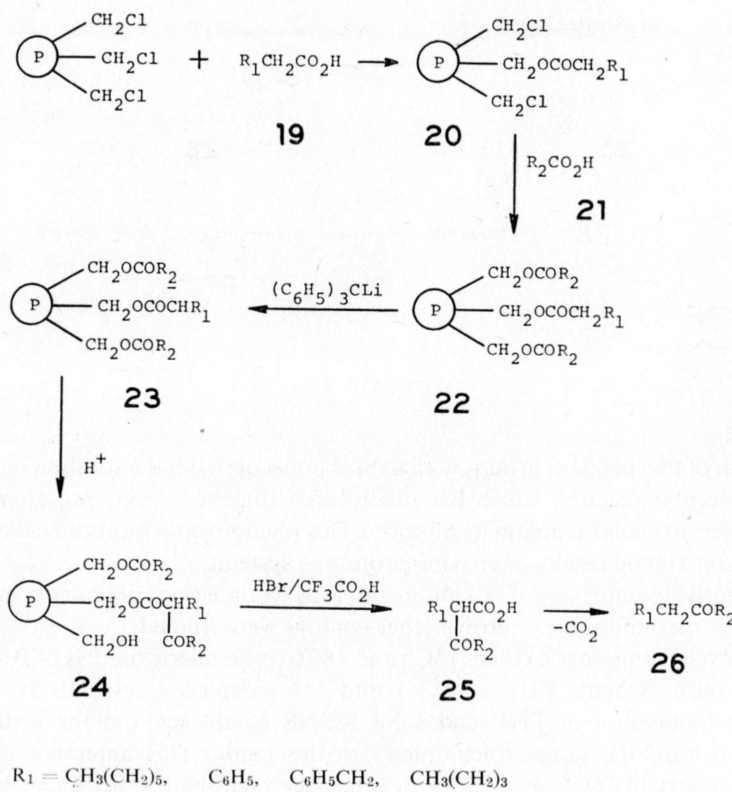

$R_1 = CH_3(CH_2)_5$, C_6H_5, $C_6H_5CH_2$, $CH_3(CH_2)_3$

$R_2 = p\text{-}ClC_6H_4$, C_6H_5

SCHEME V

A "chain doubling" reaction (Scheme VI) has been reported in the Merrifield synthesis of peptides.[8] The distance between peptide residues on **25** is such that a reaction much like that of condensation of two esters on one polymer chain occurs. This study used 0.5 to 0.7 mmole of peptide/gram of peptide resin. A lower concentration of peptide residues would tend to reduce chain doubling. Other factors that may prove to be important in these intrapolymeric reactions are the chain length of peptide, the nature of the polymeric leaving group, degree of cross-linking of the support, and the solvent system.

1.3 Cooperative effects

The fact that juxtaposition of reactive moieties can be "insured" if they are pendant groups of a polymer is the basis for another advantage of polymeric reagents which has been labeled "cooperative effects". In a polymer the

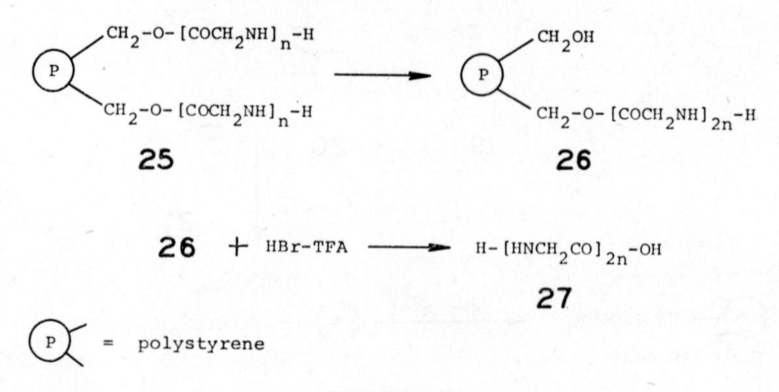

SCHEME VI

reaction of two pendant groups with a third molecule from the solution mimics a bimolecular reaction which has much less stringent entropy requirements than a termolecular reaction in solution. This phenomenon has been called on to explain kinetic results of enzyme prototype systems.

Recently, complexes of alkali metal ions with either polymeric crown ethers or their monomeric crown ether analogs were studied.[9]

The cyclic monomeric ethers 15C5 and 18C6 (monomeric models of P-15C5 and P-18C6, Scheme VII) form 2:1 and 1:1 complexes, respectively, with fluorenylpotassium in THF and solid KCNS complexes and the evidence points toward the same stoichiometry in this study. This difference in the stoichiometry of 15C5 and 18C6 ethers has been reasonably attributed to the difference in ring diameter of the two ethers.

The generally larger ratios of polymeric to monomeric complexing strength in the 15C5 ether systems compared to the same ratios in the 18C6 system

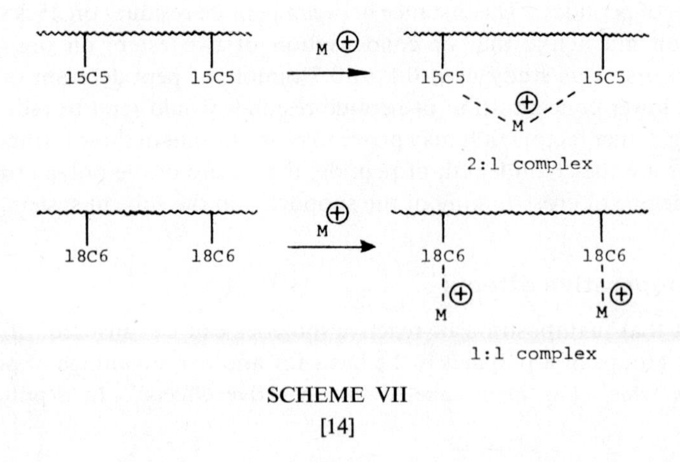

SCHEME VII

(Table IV) is expected if two crown units complex with each alkali metal ion in the former system. Two crown units are held in close proximity in the polymer and the 3 unit complex would form more readily than would the 3 unit complex in the 15C5 monomeric system. Since the complex is of a 1:1 nature in the 18C6 system the difference between polymeric and monomeric systems would not be expected to be as great.

TABLE IV

Ratios of complexing abilities of analogous polymeric *versus* monomeric crown ethers with alkali metal ions

Cation	Poly-15C5/15C5	Poly-18C6/18C6
Li^+	> 10	6.2
Na^+	2.3	1.6
K^+	3.8	1.1
Rb^+	6.4	1.1
Cs^+	19.7	

1.4 Selectivity of substrate

1.4.1 *Size* If the active site of a macromolecule is located inside a semi-ordered insoluble polymer, it is not difficult to visualize a situation in which the rate of reaction of substrates would depend on their ability to gain access to this active site. For example, the size of a molecule would certainly be one such factor.

A system such as is described above has been reported in a study in which the selective hydrogenation of olefins was demonstrated with a polymer-supported rhodium(I) catalyst.[10] As Table V shows the relative rate of reduction of cyclic olefins depends upon the size of the ring when the reducing agent is coupled to a macromolecule, **29**, while this effect is not evident with tris(triphenylphosphine)chlororhodium(I) (see also Scheme VIII).

1.4.2 *Hydrophobicity* In aqueous solutions, apolar bonding between two hydrophobic molecules has been shown to be a factor in increasing the rate of hydrolysis of esters by polymeric catalysts.

TABLE V

$>C=C<$	Beads, **29** relative rates	$RhClL_3$ 2.5 mmol
Cyclohexene	1	1.0
Cyclooctene	1/2.54	1.0
Cyclododecene	1/4.45	1/1.5
Δ^2-Cholestene	1/32	1/1.4

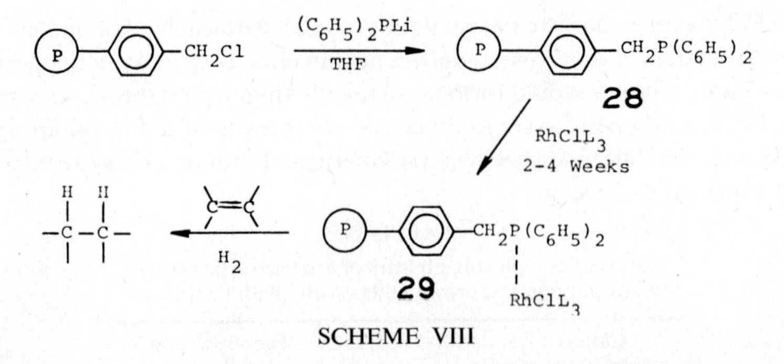

SCHEME VIII

The esterolytic action of the synthetic, macromolecular catalyst, poly-4(5)-vinylimidazole (PVIm) has been investigated.[11] The rate of the PVIm-catalyzed esterolysis of a long-chain ester (S_{12}^-) was *ca.* 10^3 times as fast as that of the monomeric imidazole (Im) catalyzed esterolysis of the same ester.[12] These esterolytic rates were found to be critically dependent on the chain length of the acid portion of the substrate and one the solvent composition. The bulk of the rate enhancement was thus attributed to apolar association of catalyst and substrate.

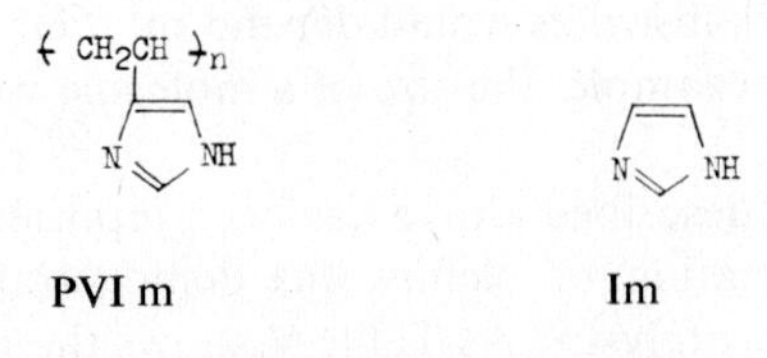

PVI m　　　　Im

During the study of the PVIm and Im-catalyzed hydrolyses of S_n^- in varying vol% ethanol–water a very interesting autocatalytic or *accelerative kinetic pattern* was observed (see Figure 1). The rate of the reaction continuously increased so that at 75% completion the rate was *ca.* 5 times faster than the initial rate.

By observing the deacylation rates of the polymeric acyl imidazoles from the hydrolysis of S_n^- (see Scheme IX), it was noted that for the cases exhibiting the accelerative behavior, deacylation was the rate-determining or slow step. This demonstrates again the influence of apolar bonding on the rates of esterolysis. The acylated polymer is more hydrophobic than the non-acylated polymer.

Table VI shows the surprisingly large dependence of the deacylation rate on the chain length of the acyl group. This observation is rationalized by considering formation of a polysoap-like structure as the polymer is increasingly acylated with long-chain groups.[13]

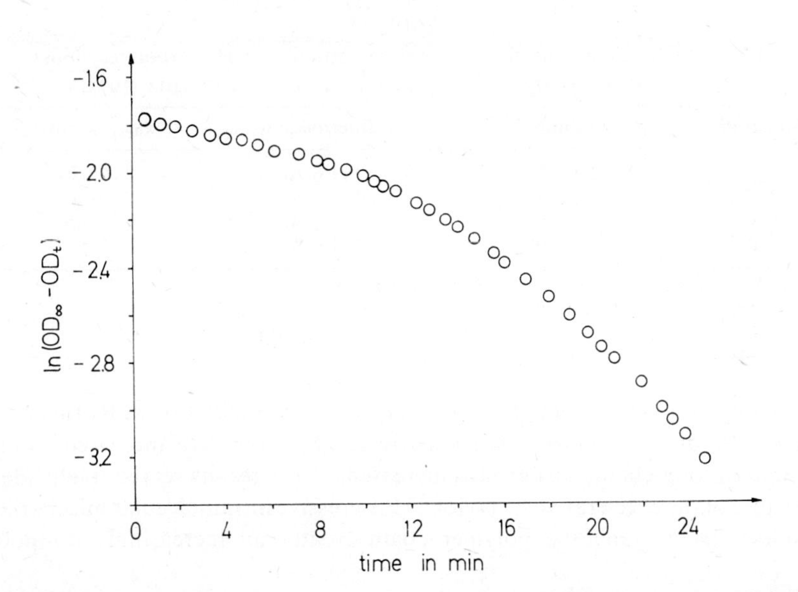

FIGURE 1 Pseudo-first-order plot of the esterolysis of cyclododecene catalyzed by poly-4(5)vinylimidazole (PVIm) shows accelerative kinetic pattern. 40 vol.perc. C_2H_5OH in H_2O, [PVIm] $= 5.0 \times 10^{-4}$ mol/dm^3, [S_{12}^-] $= 5.0 \times 10^{-5}$ mol/dm^3, $\mu = 0.02$, TRIS $= 0.02$ mol/dm^3, pH $=$ 8.0, 26°C. [Reprinted with permission from *J. Am. Chem. Soc.*, **95,** 6014 (1973).]

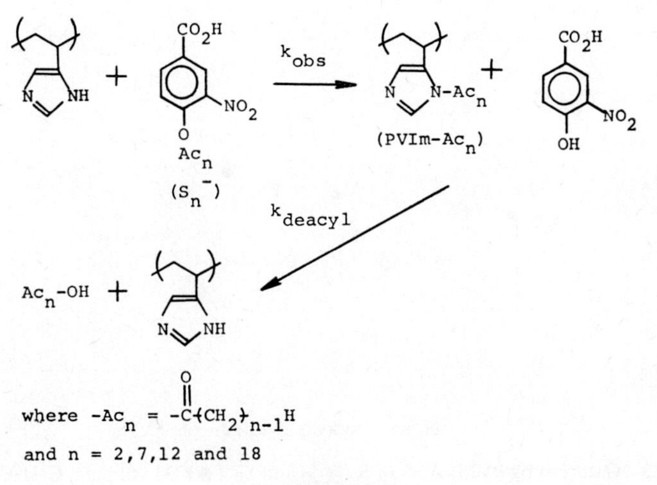

SCHEME IX

[Reprinted with permission from *J. Am. Chem. Soc.*, **95,** 6014 (1973).]

TABLE VI

First-order rate constants for hydrolysis (acylation) and deacylation reactions[a]
[Reprinted with permission from *J. Am. Chem. Soc.*, **95,** 6014 (1973).]

Substrate	$k_{obs}(min^{-1})$[b]	Intermediate	$k_{deacyl}(min^{-1})$[c]
S_2^-	0.022	PVIm-Ac_2	0.250
S_7^-	0.013	PVIm-Ac_7	0.242
S_{12}^-	0.090[d]	PVIm-Ac_{12}	0.041
S_{18}^-	0.500[d]	PVIm-Ac_{18}	0.006

[a] 40 Vol% ethanol–water, $\mu = 0.02$, pH = 8.0, 26°C.
[b] [PVIm] = 5.0×10^{-4}mol/dm³ [S_n^-] = 5.0×10^{-5}mol/dm³.
[c] Determined for >90% deacylation completed; deacylation is accelerative.
[d] Accelerative kinetic behavior, k_{obs} at *ca.* 75% reaction.

Figure 2 shows the effect of varying percent dodecanoylation of PVIm on the hydrolyses of S_7^- and S_2^-. Particularly for S_7^-, the rate increased as the content of long-chain, acyl groups increased. These results served to elucidate the cause of the accelerative behavior.[13] Progressive intramolecular micellarization occurring along the polymer chain creates an increasingly nonpolar

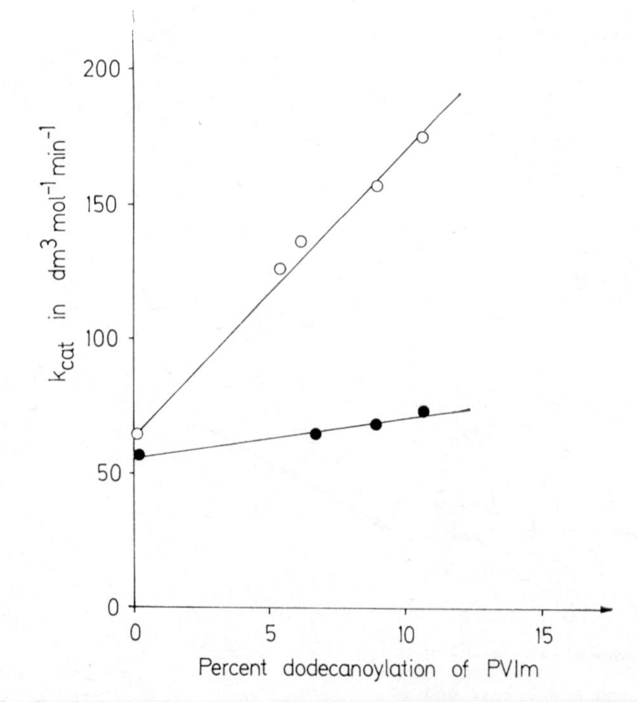

FIGURE 2 Catalysis by PVIm-Ac_{12} of S_7^- (○) and S_2^-(●). 33 vol.perc. C^2H_5OH in H_2O, $\mu = 0.02$, TRIS = 0.02 mol/dm³, pH = 8.0, 26°C. [Reprinted with permission from *J. Am. Chem. Soc.*, **95,** 6014 (1973).]

molecule. This results in an enhancement of the apolar association of substrate and catalyst. Undoubtedly, a conformational change accompanies this process of polysoap formation which is evidenced through the kinetics of the hydrolysis reaction.

The situation described above may be considered conceptually similar to the *allosteric effect* displayed by certain enzyme systems.[14] The catalytic function of enzymes may be affected and controlled by interaction with small molecules, not only directly, at the active site, but also indirectly, at distant, secondary allosteric sites. Generally, a conformational change accompanies the observation of the allosteric phenomena. The accelerative behavior is evidence of a conceptually similar occurrence in the case of the synthetic, macromolecular catalyst, PVIm.

Hydroxylamine is known to be more than 10^5 times as effective as water in acyl transfer from acetyl imidazole.[15] Figure 3 demonstrates the effect of added hydroxylamine on the accelerative behavior. The acceleration is diminished with increasing hydroxylamine content in the system; this can be explained by a decrease in the concentration of the intermediate. Eventually, the accelerative behavior is completely eliminated, indicating that the intermediate is being destroyed at a faster rate than the substrate is being hydrolyzed.

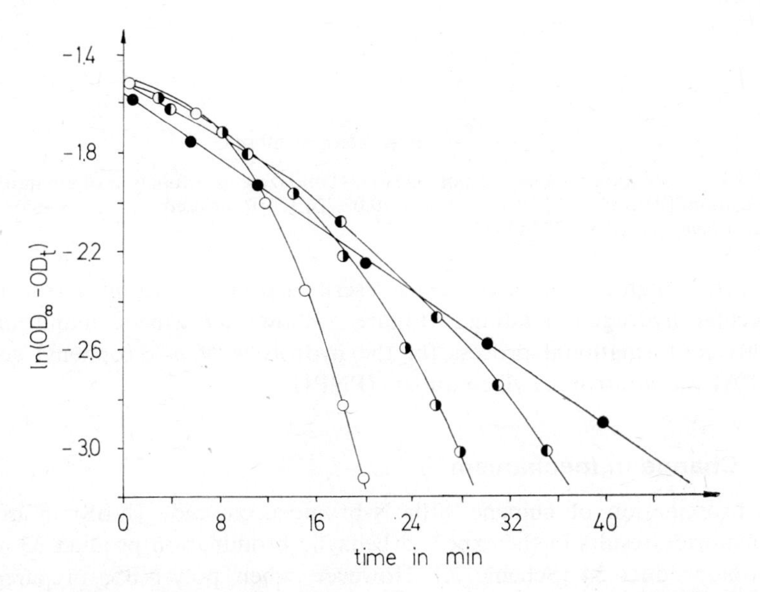

FIGURE 3 Effect of increasing concentration of hydroxylamine on accelerative behavior. 40 vol.perc. C_2H_5OH in H_2O, [PVIm] = 5.0×10^{-4} mol/dm³, [S_{12}^-] = 5.0×10^{-5} mol/dm³ (○), with [NH_2OH] = 3.0×10^{-4} mol/dm³ (◑), 4.5×10^{-4} mol/dm³ (◐), 11.1×10^{-4} mol/dm³ (●).

The viscosity of PVIm in varying vol % ethanol–water (Figure 4) indicates a contraction of the macromolecule in low and high ethanol compositions.[16] Shrinkage of the polymer coils in low ethanol content might be the result of intra- and intermolecular nonpolar interactions, whereas its shrinkage in

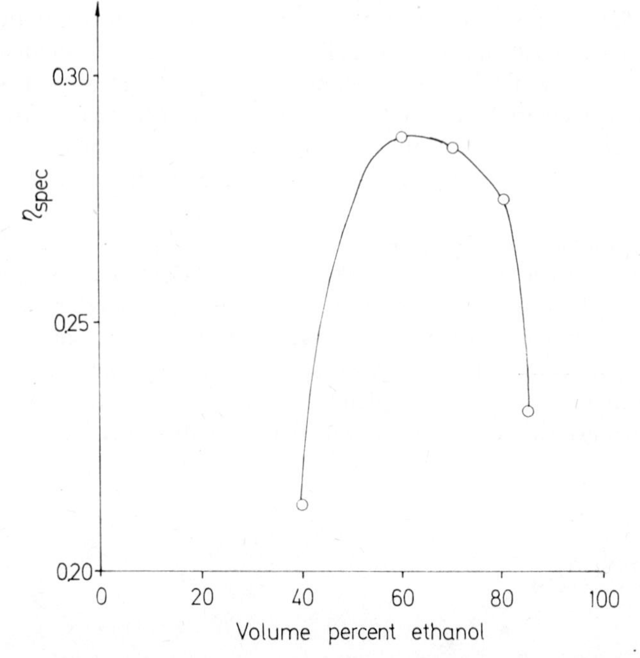

FIGURE 4 Solution viscosity of poly-4(5)-vinylimidazole as a function of ethanol/water composition. [PVIm] = 4.229 g/dm^3, μ = 0.02, 26°C. [Reprinted with permission from *J. Am. Chem. Soc.*, **93,** 3222 (1971).]

solvents of high methanol content is ascribed primarily to intra- and intermolecular hydrogen bonding.[16] Figure 5 shows the kinetic manifestation of this conformational process for the hydrolysis of *p*-nitrophenyl acetate (PNPA) and *p*-nitrophenylheptanoate (PNPH).

1.5 Change in mechanism

The bromination of cumene with N-bromosuccinimide (NBS) in carbon tetrachloride results in the expected benzylic bromination product **33** and a dibromoproduct **34** (Scheme X). However, when poly-NBS, prepared by bromination of polymaleimide, was used as the brominating agent, a completely different set of products was obtained[17] as shown in Scheme X. This demonstrates another feature of macromolecular reagents; that of a

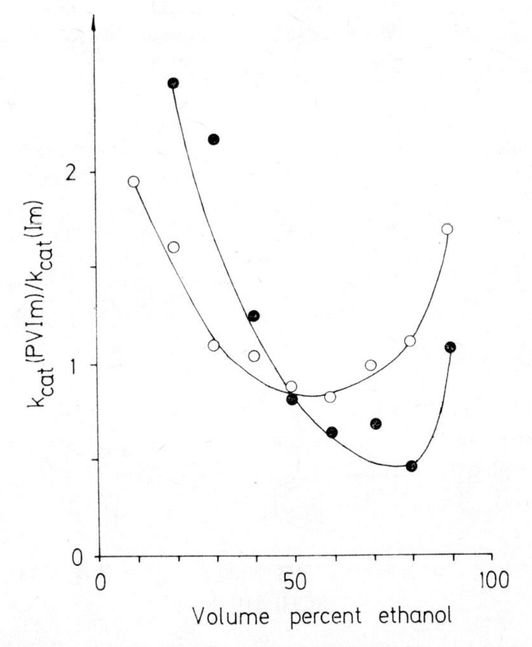

FIGURE 5 Ratios of rate constants of the poly-4(5)vinylimidazole (PVIm) and monomeric imidazole (Im) catalyzed solvolysis reactions of PNPA (○) and PNPH (●) as a function of the ethanol concentration in water at 26°C, pH $\approx$ 8, and $\mu = 0.02$. [Reprinted with permission from *J. Am. Chem. Soc.*, **93**, 3222 (1971).]

change in mechanism (Scheme XI) upon anchoring the functional moiety to a large carrier molecule.

The polar medium provided by neighboring succinimide units in the polymer may be responsible for the (especially the repeated dehydrobromination) changed mechanism and different products obtained in NBS and poly-NBS brominations. This idea is substantiated by the fact that the bromination of cumene with NBS in a *more polar* solvent, acetonitrile, results in products similar to those obtained with poly-NBS in carbon tetrachloride.

1.6 Simple separation of polymeric reagent

Polymeric phosphoranes have been prepared (Scheme XII) and reacted with carbonyl compounds to form olefins.[18] These Wittig reactions on polymeric supports afford product yields comparable to monomeric reactions but the removal of the phosphine oxide formed in the reaction is done by simply filtering the polymer to which the P–O unit (**39**) remains attached. This advantage is, in principle, one of ease in separation, much like solid phase peptide synthesis.

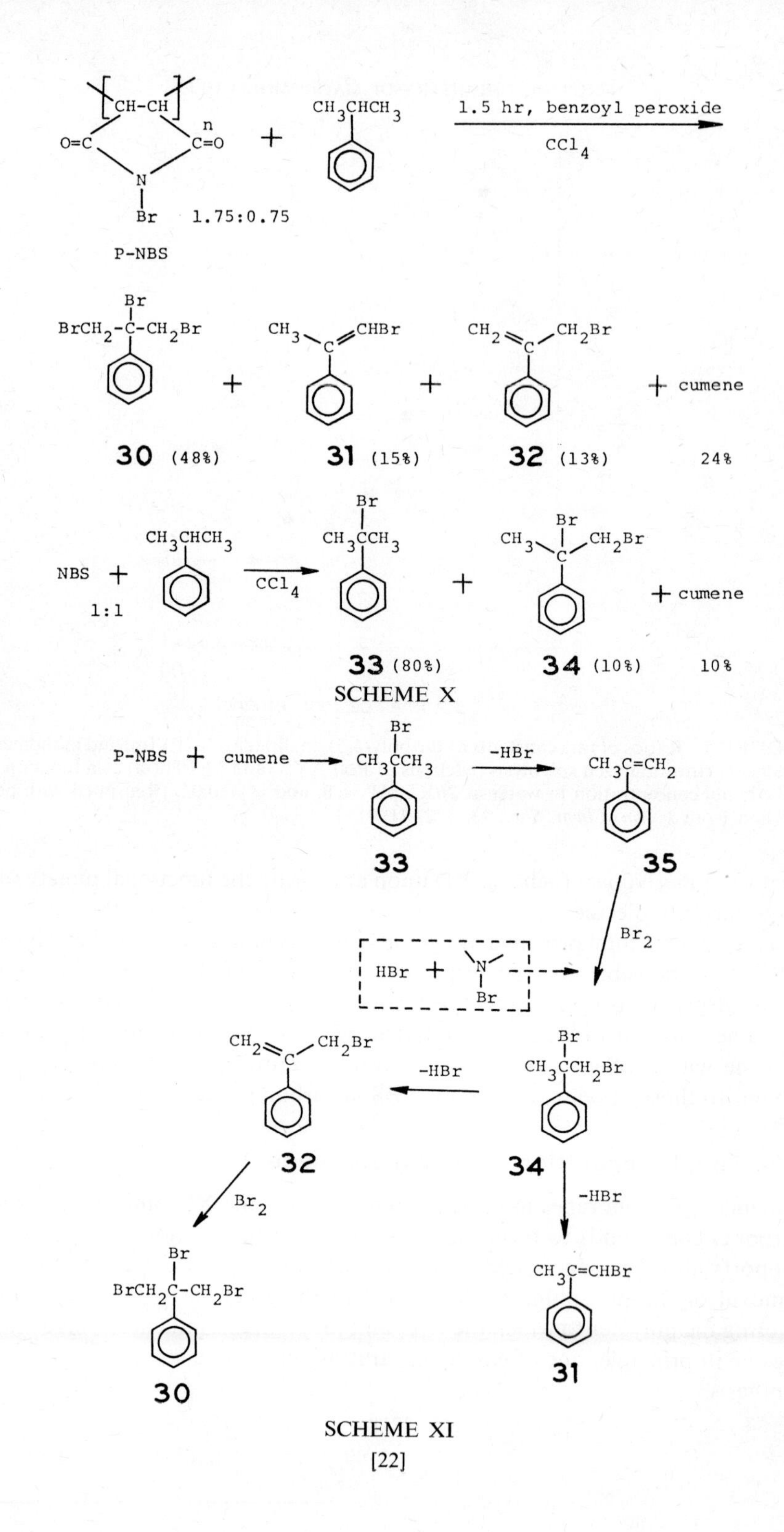

SCHEME X

SCHEME XI

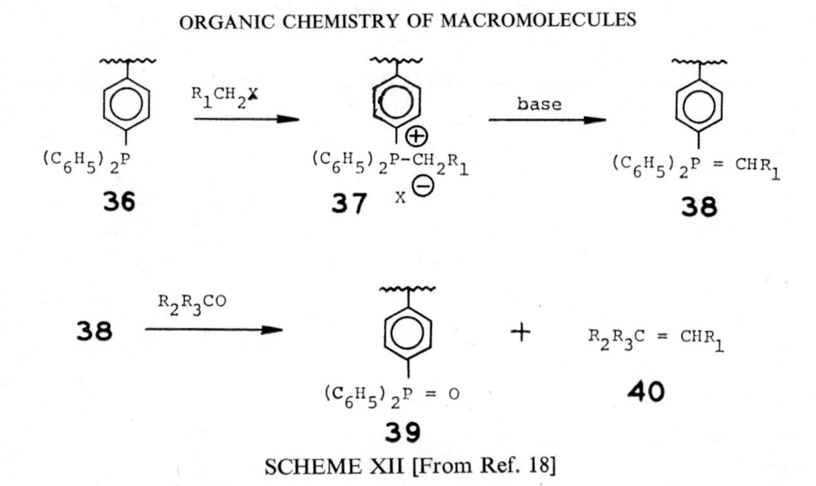

SCHEME XII [From Ref. 18]

2 HIGH TEMPERATURE POLYMERS—POLYPHENYLS

It is not my purpose here to review the subject of polymers useful at high temperatures. Polyphenyls, of course, have always been attractive since they provide a structure which does not allow a low activation energy rate process such as oxidation or hydrolysis to be important. Included here are several examples of some recent syntheses of polyphenyls designed particularly to make them more soluble in organic solvents.[19–22]

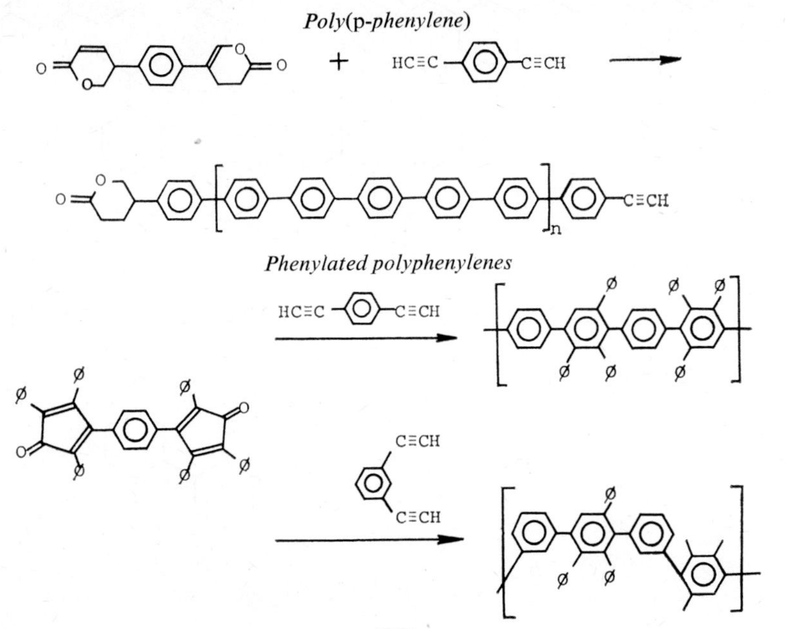

Branched chain polyphenyl

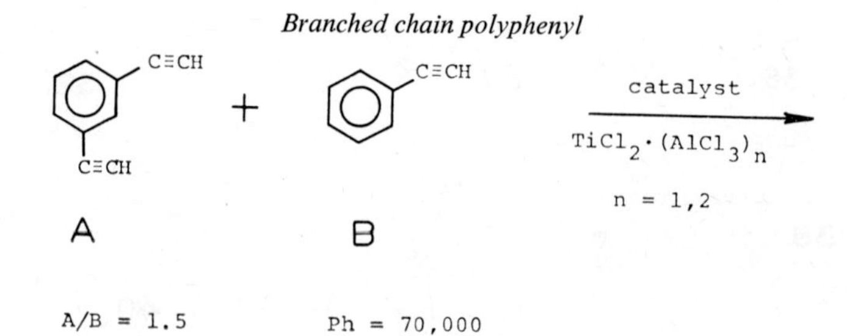

Structure of polyphenylenes [From Ref. 22]

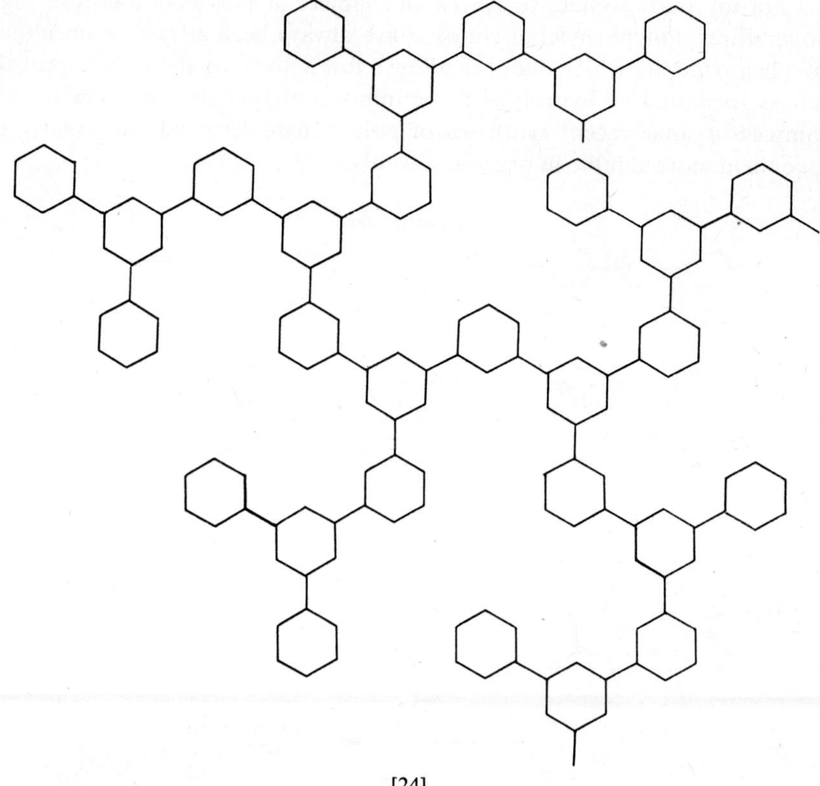

References

1. For a good account see, G. R. Stark, *Biochemical Aspects of Reactions on Solid Supports*, Academic Press, New York, N.Y., 1971, Chapter 3.
2. M. Fridkin, A. Patchornik and E. Katchalski, *J. Amer. Chem. Soc.* **87,** 4646 (1965).
3. A. Patchornik, M. Fridkin and E. Katchalski, *Proc. Eur. Peptide Symp. 8th, Noordwijk*, 91 (1967).
4. J. I. Crowley and H. Rapoport, *J. Amer. Chem. Soc.* **92,** 6363 (1970).
5. A. Patchornik and M. A. Kraus, *J. Amer. Chem. Soc.* **92,** 7587 (1970).
6. (a) F. Camps, J. Castells, M. J. Ferrando and J. Font, *Tetrahedron Lett.* 1713 (1971); (b) M. A. Kraus and A. Patchornik, *Israel J. Chem.* **9,** 269 (1971).
7. M. A. Kraus and A. Patchornik, *J. Amer. Chem. Soc.* **93,** 7325 (1971).
8. H. C. Beyerman, E. W. B. De Leer and W. van Vossen, *Chem. Commun.* 929 (1972).
9. S. Kopolow, T. E. Hogen and J. Smid, *Macromolecules* **4,** 359 (1971).
10. R. H. Grubbs and L. C. Kroll, *J. Amer. Chem. Soc.* **93,** 3062 (1971).
11. For a review, see C. G. Overberger and J. C. Salamone, *Macromolecules* **2,** 553 (1969).
12. C. G. Overberger, M. Morimoto, I. Cho and J. C. Salamone, *Macromolecules* **2,** 553 (1969).
13. C. G. Overberger and R. C. Glowaky, submitted to *J. Amer. Chem. Soc.*
14. H. R. Mahler and E. H. Cordis, *Biological Chemistry*, 2nd Ed., Harper and Row, New York, 1971, p. 299.
15. W. P. Jencks and J. Carrioulo, *J. Biol. Chem.* **234,** 1272, 1280 (1959).
16. C. G. Overberger and M. Morimoto, *J. Amer. Chem. Soc.* **93,** 3222 (1971).
17. C. Yaroslavsky, A. Patchornik and E. Katchalski, *Tetrahedron Lett.* 3629 (1970).
18. (a) S. V. McKinley and J. W. Rakshys, *Chem. Commun.* 134 (1972); (b) F. Camps, J. Castells, J. Font and F. Vela, *Tetrahedron Lett.* 1715 (1971).
19. J. K. Stille and Y. Gilliams, *Macromolecules* **4,** 515 (1971).
20. J. K. Stille and G. K. Noren, *J. Polym. Sci.*, Part B, **7,** 525 (1969).
21. A. J. Chalk and A. R. Gilbert, *J. Polym. Sci.*, A-1, **10,** 2033 (1972).
22. W. Bracke, *J. Polym. Sci.*, A-1, **10,** 2097 (1972).

Biopolymers: Origin, Chemistry and Biology†

MELVIN CALVIN

Laboratory of Chemical Biodynamics, University of California, Berkeley, California

A brief statement concerning the way in which biopolymers may have originated in the nonbiological world is made, including experimental evidence. This also includes a discussion of such matters as the way in which the code might have originated, that is, the relationship between polypeptides and polynucleotides as well as the secondary and tertiary structure resulting from the primary structure determination. The importance of the interaction of biopolymers with lipids for the formation of limiting membranes is discussed, leading to the formation of cells and other self-organizing cellular type organelles. Thus, the second critical physical–chemical problem for cellular organization, namely, the biopolymer–lipid interaction, is now coming under scrutiny, both in terms of synthetic systems as well as natural ones.

It seems that Dr. Elias has a particular talent for asking his guest speakers critical questions. Just as his question to Professor Overberger induced Professor Overberger to reorganize his thoughts in a new framework, so his question to me has done the same thing. I constructed this collection of ideas over a period of several months, and came out with an outline which I will place on the board, so as I go through it you will see the way in which the relationships which I will try to describe are related to each other and to what Professor Overberger has just said and to what the speakers later today will continue. The *first section* of this discussion will be the origin of biopolymers, particularly of proteins, including the chemistry which may be involved. The *second part* will describe the secondary-, tertiary-, quaternary-structure of biopolymers and the interaction between proteins and nucleic acids.

†The work described in this paper was sponsored by the U.S. Atomic Energy Commission and by the Elsa U. Pardee Foundation for Cancer Research. This is a transcription of an address presented at the dedication of the Midland Macromolecular Institute, Midland, Michigan, September 29, 1972.

Following this is the *third section* on the origin of the code; the *fourth section* deals with the interaction between the proteins and lipids.

Why have I chosen this particular method of organization? Many of the phenomena (reactions) which I will discuss are actually extrapolations and modifications of reactions with which polymer chemists are familiar, but, as Professor Flory mentioned yesterday, there is a kind of curtain which keeps the polymer chemist, whose background is oriented to the structure and synthesis of macromolecules of any kind and the determination of the basic principles involved, and the group that has grown up in biology and biochemistry and which has learned about the physical chemistry and the chemistry of the behavior of the kinds of materials they deal with from quite a different point of view. The Biochemists have felt that these things are almost magic, and they believe new rules, or laws, or chemistry, may be required to understand in detail how the naturally-occurring biopolymers (proteins, nucleic acids, etc.) perform their function. There appears to be some "mystique" about this, and I hope that when I finish this discussion you will see some of the reasons for the "mystique" which is generally in the minds of those who grew up from the side of biology. I have tried to interpret the behavior of biopolymers in terms which are thoroughly familiar to you, as polymer chemists, and may even be naïve in your terminology. You must remember that most of this work is being done by people who have not come from your background, and that is why some of the material may seem obvious to you. Some of it, however, may not be obvious to you and I hope will give you suggestions of what to do next.

FORMATION OF BIOPOLYMERS, PARTICULARLY PROTEINS

Let us begin with a consideration of the general scheme by which the two major biopolymers are constructed by living organisms. These are the proteins and the nucleic acids, and a sequence is shown schematically in Figure 1, beginning with the DNA double helix of the cell which can in some way (which is still not clarified) reproduce itself. That is, it can copy a particular sequence of nucleotide monomers in some particular order, and make a complementary copy, giving rise to a double set. However, in addition to that, there is another type of reaction which occurs inside the cell. Instead of copying the DNA into another DNA strand, part of the DNA (which is an enormously long polymer, made up of essentially four bases hooked together by ribose phosphate linkages) can be copied into a ribose phosphate polymer, with a series of slightly different bases but related to the initial sequence. That part which is so copied contains in it the "message" for a particular protein. That is why this type of material is called "messenger RNA". The messenger RNA then comes out of the nucleus and thus becomes the message for constructing a particular protein. The message is contained in the sequence of bases in the polymer. As you know, it takes three of these bases to designate a particular amino acid. The particular "messages" are then hooked onto the catalysts, which are themselves made up partly of protein and partly of nucleic acid and some lipid substances, and the situation is now ripe to put together the amino acids into a peptide. In order to do that, the amino acids must be suitably prepared. Each one of these different length straight lines in Figure 1 represents a different amino acid. The wiggle line represents the amino acid, acylated with a phosphate residue (amino acyl phosphate or amino acyl adenylate). This, then, is handed over to a small piece of polymer (RNA) constructed of about 70–100 monomeric units, containing in it somewhere a certain triplet (three bases in a row) which is characteristic of a particular amino acid; this distinction is designated by the different symbols in Figure 1, representing a different sequence of the three bases (circles, lines, crosses, squares). Thus, each activated amino acid is loaded onto the terminus of a specific small oligomer (small polymer of the order of 70–100 units). We now have the amino acid loaded onto a particular transfer RNA (tRNA) with a particular triplet of bases characteristic of that amino acid. These now are free to attach themselves to the messenger RNA, which will be done at points in the messenger which are complementary to the three bases of which they are composed, thus providing the order of the amino acids which go into the polymer to make the polypeptide. The polypeptide, then, is "zipped up" to construct the protein. The protein is, in general, itself a characteristic polypeptide, have a characteristic shape and structure, which will be discussed later.

All of these reactions—the copying reaction (replication), the transcription into RNA, and the activation, the loading reaction, and the sipping up on the ribosomes—are dependent for their specificity in part at least on the proteins which are themselves thus made. This is a "chicken and an egg" problem today, and it sets the stage for the next part of the discussion.

How did this reflexive system, which is represented diagramatically in Figure 1, arise? As chemists, we would like to see if chemistry could have designed such a system and put it together in just this way. This is, of course, what I mean by the nonbiological origin of the polymers and what I mean here by the origin of the code, How did this relationship between particular triplets and particular amino acids come into being?

I believe this relationship is the result of chemistry and not a "frozen accident" as some of the biologists believe. The evidence for this is yet, however, to be forthcoming in the laboratory. I think that most of you as polymer chemists will find this a rather amenable point of view, and you will, I hope, turn to trying to devise experiments which will show there is some reason

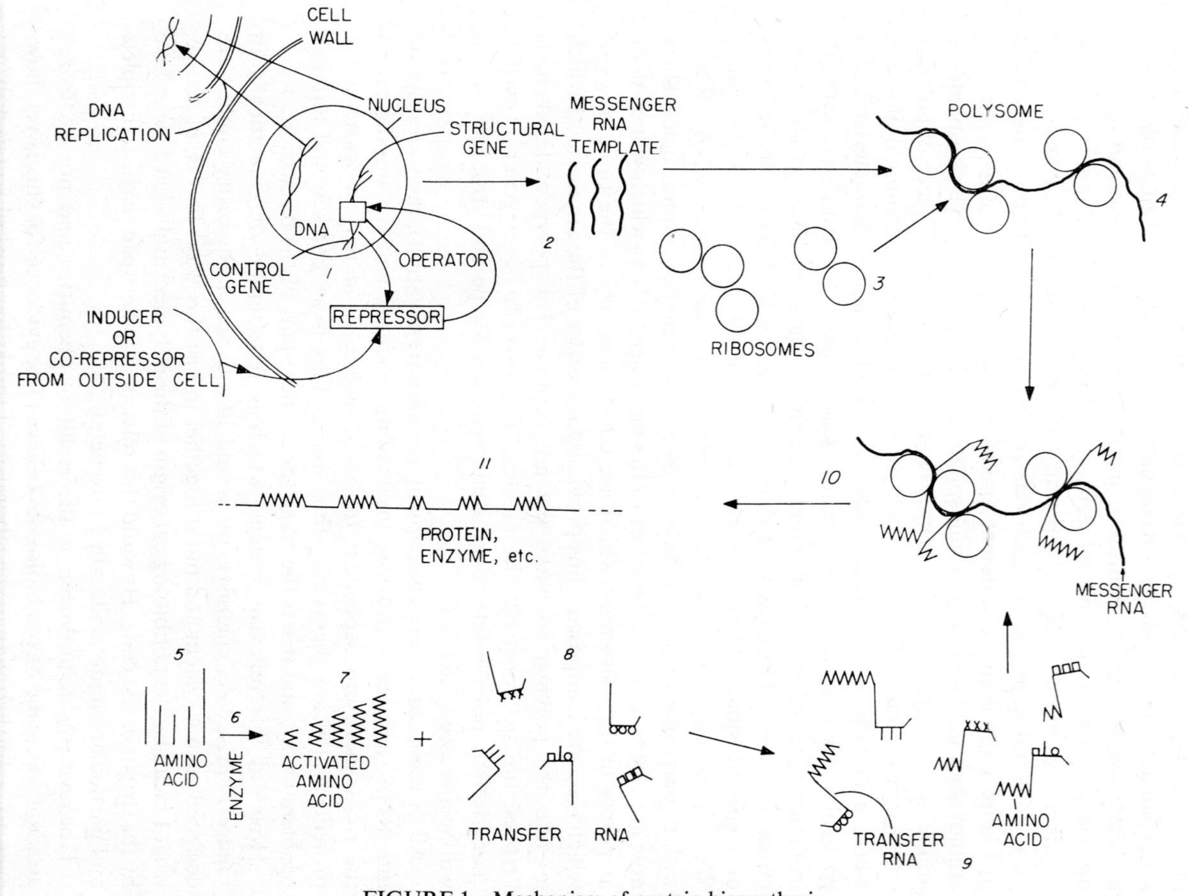

FIGURE 1 Mechanism of protein biosynthesis

for this triplet corresponding to this particular amino acid and not to some other one. That is something which remains yet to be definitely shown in the laboratory and is a subject of considerable controversy.

Returning then to possible ways in which the proteins and nucleic acids might have come into being in a nonliving world: The world with which we deal is a world in water; it is not a world in methylene chloride, dioxane, DMF, or other solvents. And all of these reactions which I have described to you in Figure 1 occur in water. I feel that whatever reactions we use, whatever reactions we call upon, to do the originating of the protein and the nucleic acid should be essentially in an aqueous medium to begin with. This can be modified somewhat by the resulting polymer itself in terms of the protein-lipid interaction. The reactions do, however, occur primarily in an aqueous environment.

We set out to try and generate polypeptides and other polypolymers in an aqueous environment.[1] Chemically and thermodynamically this is rather a difficult operation. The formation of the polymer from the monomer of each one of the biopolymers I have mentioned (proteins, nucleic acids, carbohydrates, lipids) is the result of a dehydration condensation. To make a dehydration condensation occur in water is a tricky operation. Living organisms have long ago learned how to do this, but the question is could we, as chemists, do something similar. The reactions which must be accomplished in this aqueous environment are shown in Figure 2. You can see that the formation of proteins and peptides makes use of a bifunctional molecule, and from that molecule a water molecule can be eliminated, leaving a bifunctional dimer which, in turn, can grow at either end. Formally, the same kind of system occurs in each of the four cases—proteins, polysaccharides, lipids, nucleic acids. In each case, the water molecule is eliminated to result in another bifunctional molecule. This type of operation can go on continuously. With the lipids, however, this same type of reaction is not subject to as extensive extension as with the proteins or polysaccharides; with the elimination of water the ester bond is formed, which can go on. This is, however, not quite in the same category but it is possible to make fairly large molecules. I discuss this now because the lipids play an important role in the whole evolutionary scheme.

Figure 3 shows the methods for the formation of the nucleic acids by dehydration condensation reactions. Here, actually, there are three points of dehydration involved. The first involves the formation of the amino glycoside on a base, using the NH group of the base with the glycosidic semiacetal hydroxyl of the polysaccharide. The second is the formation of the phosphate ester with the primary alcohol, forming the terminal 5′-phosphate. In this case the ribose sugar is shown which forms RNA. (The DNA, of course, is the material in which the 2′-carbon lacks the hydroxyl, i.e., carries two

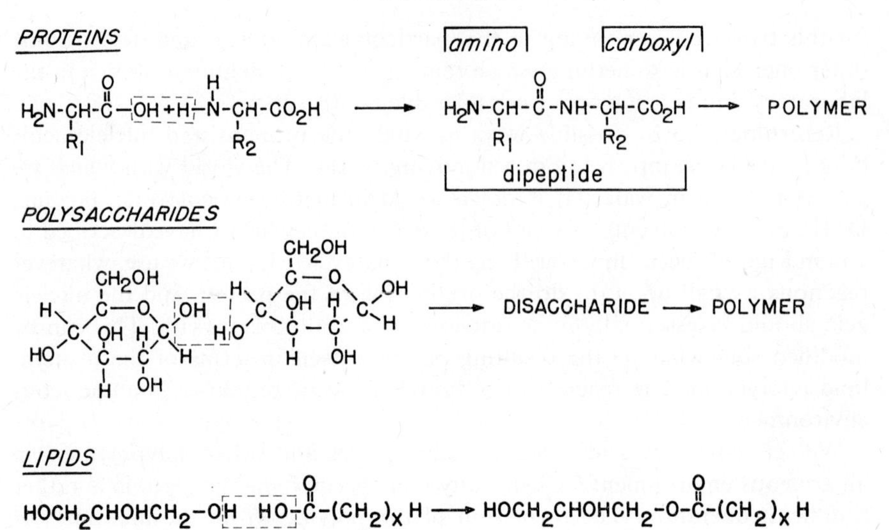

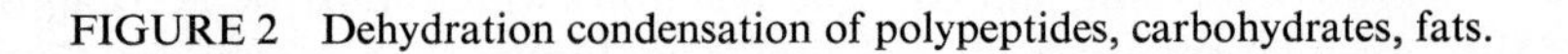
FIGURE 2 Dehydration condensation of polypeptides, carbohydrates, fats.

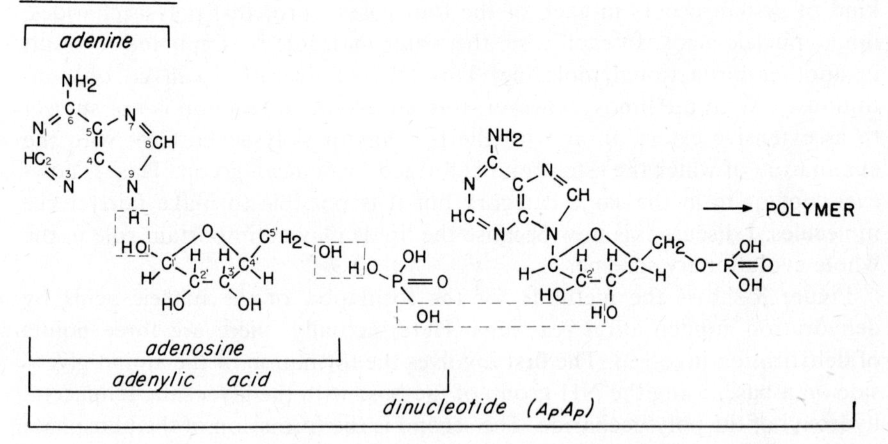

FIGURE 3 Dehydration condensation of polynucleotides.

hydrogen atoms). The third dehydration condensation occurs between the monophosphate ester and the secondary alcohol on the 3′-carbon of another nucleotide. The result is a bifunctional molecule, with both functions available; at one end the phosphate ester is available and at the other end the 3′-alcohol is usable. The nucleic acids are built on the same principles as the two other major biopolymers shown in Figure 2.

We asked the question: Is there any way to generate the polypeptides? There are many reactions for the construction of polypeptide linkages which would work in an aqueous environment. This required only a small modification of existing knowledge of reactions. One of the important reagents that has been used for the dehydration condensation of a variety of functional groups is the carbodiimide (R—N=C=N—R). The carbodiimide structure contains in it the ability to remove water, in stages, by more or less specific reactions with acid groups such as carboxyl, phosphate, and certain alcohols. This leads to condensation reactions.[2,3] This particular reagent (carbodiimide) has been used widely for nucleotide and polypeptide synthesis, but generally not in aqueous environments. We wanted to change the R groups, to make them useful in aqueous environments. The R groups can be made hydrophilic so that the carbodiimide can be dissolved in water. This is useful since the carbodiimide does not hydrolyze very rapidly directly but does so by virtue of its reaction with the carboxyl group and then with the phosphate.

The next question arose: How could this type of reaction occur in a natural environment? It turned out that there was a very easy way in which this particular structure could have evolved in a natural environment. As you may recall, ammonia was one of the primary molecules on the earth's surface, and HCN as well. With reactions among these two materials it is possible to make cyanamide, probably through cyanic acid or hydroxylamine. This is a tautomeric form of a carbodiimide. The carbodiimide in that tautomeric form is not stable; it tends to dimerize, making the dicyandiamide (DCDA) which is the common form for cyanamide. It also contains the same type of pi-bonds as carbodiimide. We used the DCDA as one of the starting materials to demonstrate this type of dehydration condensation reaction for biopolymeric materials. We took an amino acid, adjusted the pH to slightly acid conditions, and added the water-soluble carbodiimide material (dicyandiamide) to see if we could produce polypeptides in the homogeneous aqueous environment. The results of some of these experiments are shown in Figure 4.[4] We started with glycine; the reaction was performed about pH 3.5, the glycine disappears, and the dimer (diglycine), trimer (triglycine) and tetramer (tetraglycine) come out. This is a very slow reaction, and it does not go very far. You can hardly call the tetraglycine a "polymer", but at least the reaction does occur in aqueous conditions.

The next question was: Is there any selectivity of this reaction? Do the

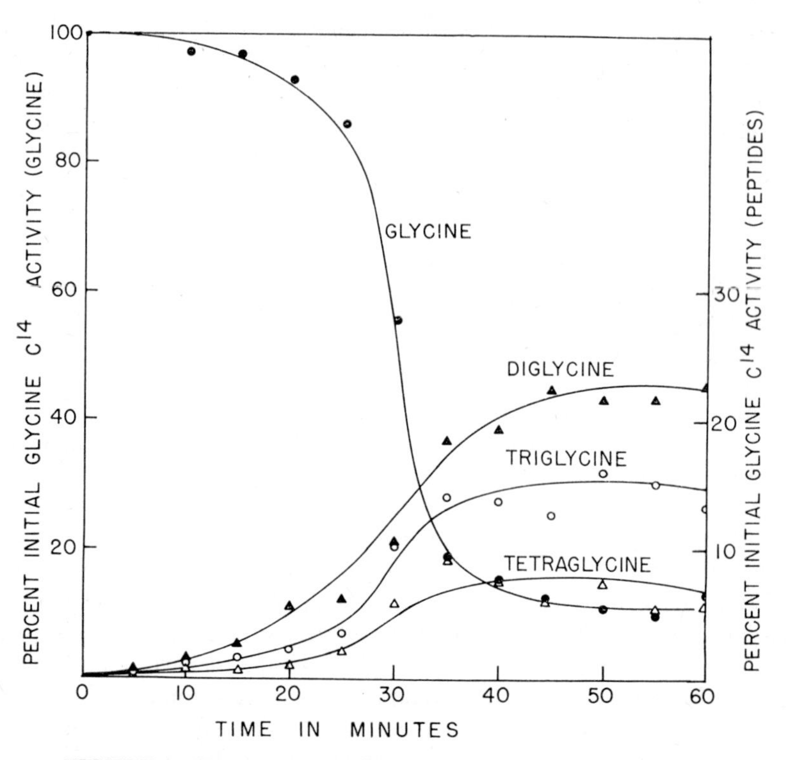

FIGURE 4 Homogeneous polypeptide formation by dicyandiamide.

amino acids self-select each other when they condense? This is a natural question to ask in the ordinary course of evolutionary studies. About five or six years ago, Steinman, in trying to answer this question put an amino acid (B) on polymer beads, through the carboxyl group and used an amino protected amino acid (A) to couple to it. He measured the relative rates of coupling of amino acid (A) in the homogeneous environment to amino acid (B) attached to the bead; this is a model, of sorts, of polypeptide formation. Table I shows the relative rates of coupling in this series, and you can see that the rates vary over a factor of ten.[5] This is the measured coupling efficiency, and the calculation is based upon the frequency of a particular pair in existing proteins, (the frequency of phenylalylglycine, glycylphenylalanine, or isoleucylglycine, etc.) as they exist in today's proteins with a statistical analysis of the situation, normalizing the frequencies to glycylglycine as unity. There seems to be slight relationship between what was found for the efficiency of coupling in the heterogeneous system and the occurrence in nature of these particular pairs of peptides. This idea has been developed still further, and

TABLE I

Comparison of experimentally determined dipeptide yields and frequencies calculated from known protein sequences

Dipeptide[a]	Values (relative to Gly–Gly) Experimental	Calculated
Gly–Gly	1.0	1.0
Gly–Ala	0.8	0.7
Ala–Gly	0.8	0.6
Ala–Ala	0.7	0.6
Gly–Val	0.5	0.2
Val–Gly	0.5	0.3
Gly–Leu	0.5	0.3
Leu–Gly	0.5	0.2
Gly–Ile	0.3	0.1
Ile–Gly	0.3	0.1
Gly–Phe	0.1	0.1
Phe–Gly	0.1	0.1

[a]The dipeptides are listed in terms of increasing volume of the side chains of the constituent residues. Gly, glycine; Ala, alanine; Val, valine; Leu, leucine; Ile, isoleucine; Phe, phenylalanine. Example: Gly–Ala: glycylalanine.[5]

there are now better ways of studying the possible self-selection of amino acids in peptide formation.

There are two other methods by which polypeptides have been made, one of which is in a nonaqueous environment. Table II shows the results of the experiments of Sidney Fox in which he took molten glutamic acid, containing a small amount of pyrophosphoric acid, as a solvent (hardly an aqueous environment) and put in an equimolar mixture of all the other amino acids, other than aspartic and glutamic.[6] Here, again, you can see that the incorporation into the resulting polypeptides is not statistical. The point of these data is to show the variabilities of occurrence, ranging from 0.5% to 5%, and the selectivity is evident.

The final study to which I want to call your attention is, perhaps, the most elegant, and I don't know whether it will continue. This is the work of Aharon Katchalsky using amino acyl adenylates as his starting point and using montmorillonite clay as the catalyst, both of which are easily available in nature. The amino acyl adenylate is available in nature because the adenylic acid can be generated by ordinary chemical reactions and the coupling of the amino acid to the adenylic acid, linking a phosphate and a carboxyl anhydride, can be easily achieved using the carbodiimide dehydration condensation reaction in aqueous environment. So the system with which Katchalsky began is achievable biologically. I do want to describe Katchalsky's discovery because it is the beginning of a very interesting idea. The amino acyl adenylate

TABLE II

Amino acid composition of proteinoid prepared with 200 ppa[6]

Amino acid	%	Amino acid	%
Thr	0.55	Lys	2.79
Ser	0.63	His	2.53
Gly	2.93	Arg	1.83
Ala	1.31	Total	7.15
Val	1.33		
Mct	0.86	Asp	51.9
Iso	0.71	Glu	13.3
Leu	3.44	Total	65.2
Tyr	3.87		
Phe	5.87		
NH_2	5.02		
Total	27.6		

Asp: Glu: equimolar mixture of basic and neutral amino acid; ratio 2:1:3; i.e., 33%, 16%, 50%; 100°C for 150 hrs. ppa = polyphosphoric acid in ppm.

used by Katchalsky is the activated amino acid in the first stage of normal biological amino acid activation; it is usually a part of an enzyme complex but, nevertheless, it is a good beginning for this type of study. He first tried to polymerize this material in an aqueous environment in a homogeneous solution. The amino acyl adenylate has in it the mixed anhydride of the carboxyl group and the phosphoric acid. It also has the free amino group of the amino acid itself. Thus, the free amino group can act as a nucleophile on the carboxyl group of the carboxy anhydride of the amino acyl adenylate.

Figure 5 encompasses the basic idea that Katchalsky introduced, but it shows more as well. If you look at the amino acyl adenylate skeleton on the polynucleotide chain you can see that the amino group can act as a nucleophile on the acyl group of the anhydride (reaction a) which would lead to the formation of a polypeptide (poly)adenylate and free adenylic acid. The alternative reaction, which Katchalsky considered but had not published on with any experimental information, is that in which the secondary hydroxyl of the sugar will attack, again, as a nucleophile on the phosphate of the phosphate anhydride (reaction b), resulting in the reverse type of polymerization, the products being a free amino acid and the (poly)amino acyl polyadenylate. There are, of course, alternative reactions which can occur (rearrangements, etc.) to stop or slow down the polymerization reactions.

What Katchalsky found in his report of 1970[7] and which he reported in more detail last May in Göttingen just before he died is that in homogeneous solution the reaction is slow and the polymers which are formed are not very large (8–10 units). He found, however, that if he performed this reaction

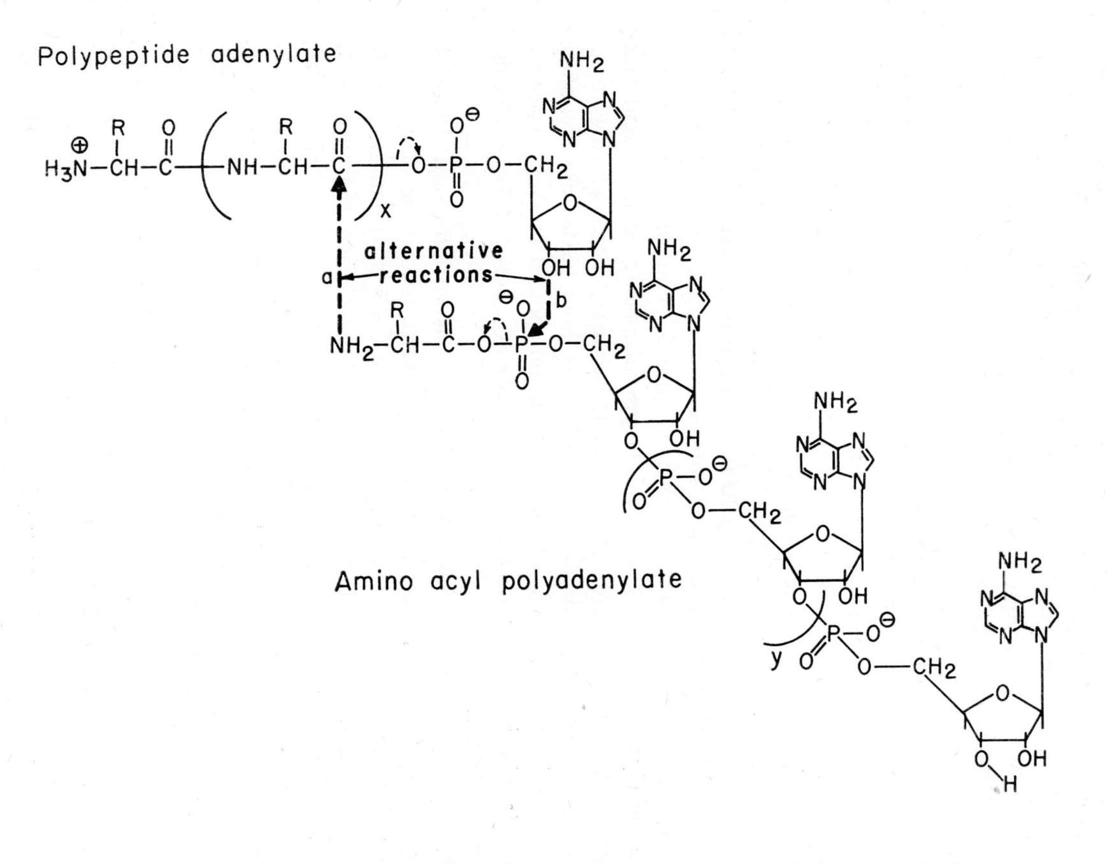

FIGURE 5 Polymerization of amino acyl adenylate, alternative reactions (Katchalsky).

in the presence of a properly prepared montmorillonite, the whole system changed. The reaction became extremely rapid and the polypeptides went to fairly high degrees of polymerization—high for this kind of a reaction, that is.[8]

We have attempted this same type of reaction, and indeed, he and I were moving in this same direction, making both polypeptide and polynucleotide from the same starting material. We can now begin to see the real relationship between the polynucleotide and polypeptide, or the nucleic acid and the protein, and how they can arise as a result of the stereochemistry of the reaction.

Katchalsky found that there were a number of discrete polypeptides formed. The result is given in Figure 6, which is a chromatogram of a second fraction. Table III also gives more information on the two principal components which were in the initial extract of this reaction. Fraction I is mostly the adenylic acid and three groups of polypeptides. This was a one gram reaction which resulted in 450 mg of adenylic acid, 20 mg of a 1000 MW substance, and another 20 mg of a 2000 MW substance. Fraction II, however, contained mostly polypeptides with some adenylic acid. The gel permeation chromatography of fraction II is shown in Figure 6.[7] You can see the presence of some eight components, with the largest one being adenylic acid, and components 1, 2, 3 and 4 are polypeptides; there are two peaks which were not identified which were inorganic salts and two more peaks, not identified, were organic in nature. When I spoke to Aharon about it he said he thought they might be polynucleotides, and I think they are; however, they have not yet been identified.

Perhaps in the future someone else will carry this work forward: I think it represents an extremely important development in the nature of the polymerization reaction. Why is it that you do not get a distribution; why do you get discrete groups of polymers? I think these discrete groups occur because it is not a single addition reaction. Some amino acids are being added, and some polynucleotides are being constructed to make a polypeptide-polyadenylate, and these materials are regulating the size of the polypeptides which are found. It will probably turn out that the polynucleotide is also discrete in some fashion.

THE ORIGIN OF THE CODE

As perhaps some of you are aware, in today's living organisms the specification of the linear array of amino acids in a polypeptide is contained in a corresponding linear array of bases in a polynucleotide. A series of three bases in the polynucleotide (DNA) specifies a specific amino acid in a polypeptide, or protein. This correspondence of a triplet of bases with a particular amino

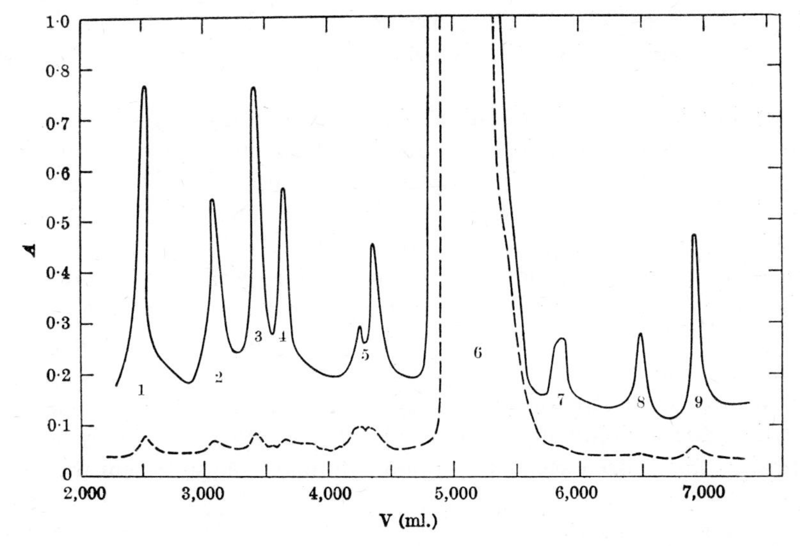

Chromatography of fraction II on 'Sephadex G-25', with gel bed 3 × 4 × 150 cm.

FIGURE 6 Polypeptides resulting from alanyl adenylate on montmorillonite (Katchalsky).

TABLE III

DATA ON PEPTIDES I

Peptide Mol. Wt.	Degree of Polymerization	Wt. (mg)
640	9	10.3
1,120	16	35.2
1,900	27	20.0
Adenylic acid	1	470.5

PEPTIDES II OBTAINED ON POLYCONDENSATION OF ALANYL ADENYLATE

Peak No.	Mol. Wt. of Peptide Adenylate	Degree of Polymerization	Wt. (mg)
4	2,130	30	8
2	2,310	32	5.2
1	3,020	42	10.8
3	4,000	56	17.1
6	Adenylic acid	1	260.1

acid is universal in all living organisms. One form of expressing this is shown in Table IV. Here you can see there is a certain amount of redundancy in the code, but the universality of it throughout the living world is now fairly well established.

Thus, in addition to the question of how these biopolymers may actually have been formed abiogenically we must address ourselves to the question of how the linear relationship between the two major biopolymers, that is, the polypeptide (protein) and the polynucleotide (DNA), evolved. That is: What is the possible origin of the code?

There are those who believe that the particular specificity we now find in living organisms between a particular triplet and a particular amino acid as assembled in Table IV is the result of an accident in some early catalytic reaction which, by virtue of a selective advantage in producing autocatalytic systems, has now been frozen into *all* of biology.[9] It is selected because a single autocatalytic system eventually dominated all the others. Another way of expressing this "frozen accident" notion is to make the statement that given

TABLE IV

SECOND LETTER	FIRST LETTER: A		C		G		U	
A	AAA AAG	Lys	CAA CAG	Gln	GAA GAG	Glu	UAA UAG	Termi- nate
	AAC AAU	Asn	CAC CAU	His	GAC GAU	Asp	UAC UAU	Tyr
C	ACA ACG ACC ACU	Thr	CCA CCG CCC CCU	Pro	GCA GCG GCC GCU	Ala	UCA UCG UCC UCU	Ser
G	AGA AGG	Arg	CGA CGG CGC CGU	Arg	GGA GGG GGC GGU	Gly	UGA UGG	Term. Trp
	AGC AGU	Ser					UGC UGU	Cys
U	AUA AUG	Ilu Met	CUA CUG CUC CUU	Leu	GUA GUG GUC GUU	Val	UUA UUG	Leu
	AUC AUU	Ilu					UUC UUU	Phe

exactly the same starting conditions and allowing chemical evolution to occur once again, that a different codeal relationship might very well have appeared.[10,11] It is my personal belief, however, that this would *not* be the case. I believe that the codal relationship reflects some characteristic molecular interactions between amino acids (polynucleotides) and nucleotides (polynucleotides).

A modicum of evidence for this already exists in the form of some experiments done on the rate of coupling of amino acids to nucleotides in a specific instance. Some years ago[12] we attached a nucleotide (in this case adenylic acid) to a synthetic polymer (in this case polystyrene). The polystyrene was in the form of microspheres. The next step was to measure the rate of coupling of two different amino acids to each of these two polystyrene-nucleotide preparations. The coupling was performed using an N-protected amino acid adenylate. The reaction is shown in Figure 7 and the results are given in Table V.[13] Here you can see that glycine reacts more rapidly with both nucleotides than does phenylalanine and adenine reacts more rapidly with both amino acids than does cytosine. The overall range of reactivity is roughly a factor of three, the slowest reaction being phenylalanine with cytosine and the most rapid the reaction of glycine with adenine.

Here the beginning of a kind of selectivity in a reaction rate is apparent.[14] However, I believe that this kind of selectivity will be very much enhanced if both the amino acid and the nucleotides are each, respectively, part of a polymer. For example, I would expect that the alternative reactions (a) and (b) indicated in Figure 6 will be very dependent upon the nature of the amino acid and the nature of the nucleotide involved. Furthermore, I would expect that selectivity would be increased with the length and character of the polynucleotide, on the one hand, and possibly even with the polypeptide on the other. This last experiment is yet to be done, but is underway.

SECONDARY AND TERTIARY STRUCTURE OF PROTEIN AND INTERACTION WITH NUCLEIC ACIDS

This part of our discussion will describe what is known about the secondary, tertiary and quaternary structures, mostly of proteins. There are several principles which I want to emphasize here: One is the way in which proteins fold up, in their secondary and tertiary structures, another is the way enzymes interact with substrates, and a third is the way in which the proteins interact with the nucleic acids. (Finally, I wish to bring up the protein–lipid interaction, Part III.)

Myoglobin, with its tertiary structure, is shown in Figure 8. The left-hand side of the figure shows the alpha helical structure and the drawing on the

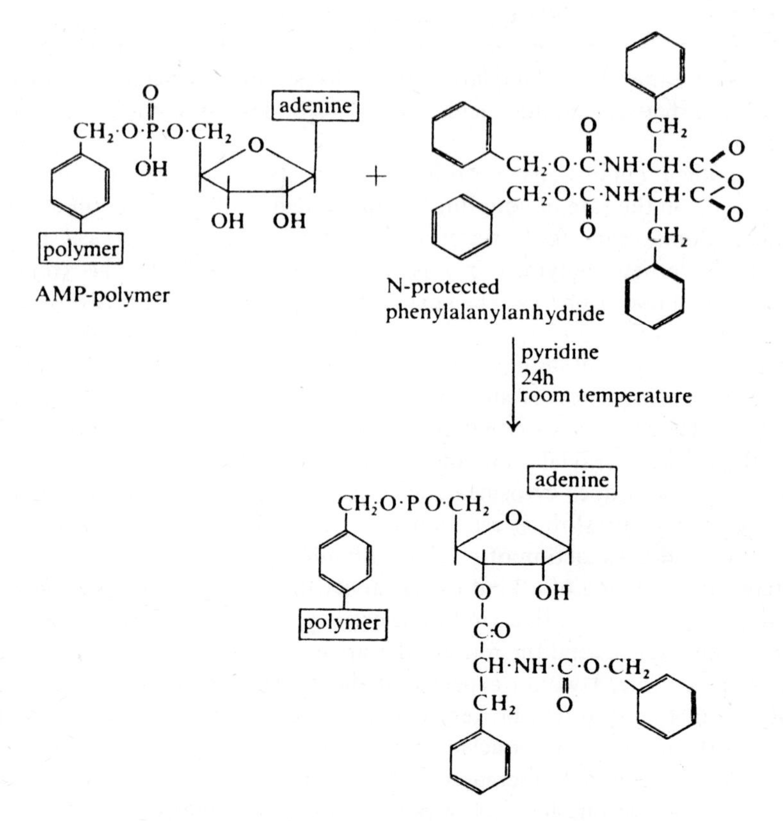

N-protected phenylalanyl -AMP- Polymer

FIGURE 7 The coupling of the polymer-AMP complex with the anhydride form of an N-protected amino acid.

TABLE V

PERCENT OF BOUND NUCLEOTIDE REACTED

Amino acid \ Base	Adenine	Cytosine
Phenylalanine	6.7	2.9
Glycine	10.0	6.5

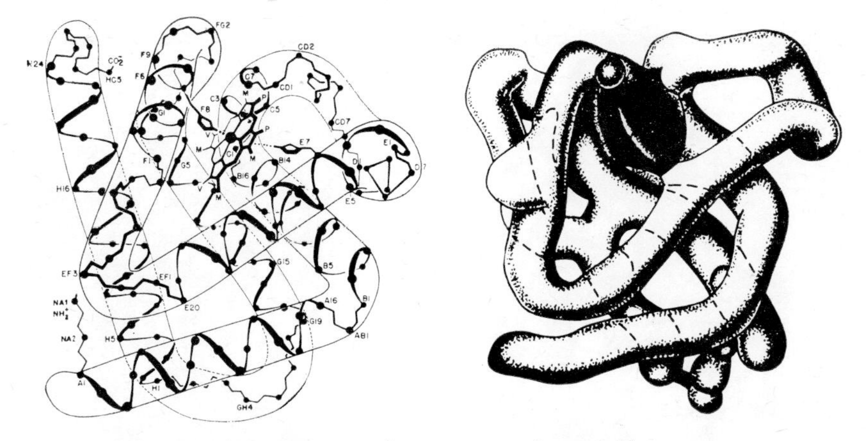

FIGURE 8 Tertiary structure of myoglobin.

right-hand gives a more three-dimensional view of the same material, to show that the myoglobin is a folded protein, with the heme stuck in the center. This secondary and tertiary structure is built into the molecule as a result of the amino acid sequence in the polypeptide which takes up this structure, given the opportunity in water.[15] The structure of cytochrome c, showing the hydrophobic internal arrangement and hydophilic external arrangement, is shown in Figure 9. The colors on this figure are classified according to the hydrophilic and hydrophobic characters of the amino acid residues. By observation, you can see that the hydrophobic parts are more or less in the middle and the hydrophilic (the cool ones) parts are on the outside of the molecule. This is the only characteristic which has so far approached a generalization on the structure of soluble proteins. In general, the structure of soluble proteins is such that when they fold up in their secondary and tertiary structures they do so in a way which places the hydrophobic chains on the inside (away from the water environment) and the hydrophilic chains on the outside, near the water environment.

It is for this reason that I had the reservation, which I mentioned earlier, about the generation of polypeptide and polynucleotide structures in a purely aqueous environment. You see that there can be modifications during the course of their construction which will remove their functions from the water environment into a nonaqueous environment, by virtue of their own structure, so to speak.

The structure of lysozyme, showing the substrate cleft, is given in Figure 11. I believe it is easy to distinguish between the lysozyme itself and the shaded area which is the synthetic disaccharide substrate which fits right into the active site cleft. The same molecule, lysozyme, is shown again in Figure 10,

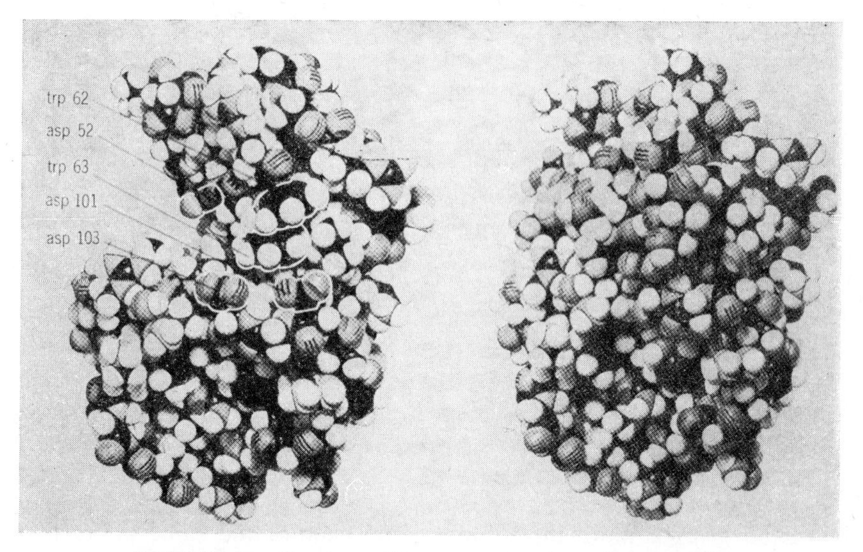

FIGURE 11 Structure of lysozyme showing substrate cleft.

illuminating the mechanism of the polysaccharide hydrolysis, and showing the actual function of the active site. You can see the polysaccharide lying in the active site of the lysozyme. The bond which is hydrolyzed is the glycosidic linkage, by means of transfer to the polymer (Asp 52) of one part of the substrate, thus letting the other part go free. The Asp glycosidic ester, in turn, is very rapidly hydrolyzed. This is the way in which the hydrolytic enzymes and many of the transferring enzymes function. Thus, the tertiary structure is evolved to give the particular tertiary architecture which will take hold of the proper substrate.

The next figure (which is an old one) reaches into the area of visibility. This is collagen (Figure 12), showing that we can refold molecules into tertiary structure but, even further than that, the molecules can actually reassemble into more complex quaternary structures which are dependent upon their primary sequence.[16] The upper part shows the separated collagen fibrils, and in the bottom sequence are the reaggregated fibrils which appear similar to the natural ones. The fibrils are highly ordered arrays and this is a protein–protein aggregation. The next few figures show the dissociation and re-aggregation of the tobacco mosaic virus (a relatively simple virus particle) which consists of a set of identical protein molecules and one nucleic acid molecule. These two can be separated and then reassembled again, into a complete virus particle.[12] Since that time, more complex virus particles have been reconstituted as well. Figure 13a shows the native TMV virus, with the uniform particles, and Figure 13b shows the reassembly of the TMV protein itself

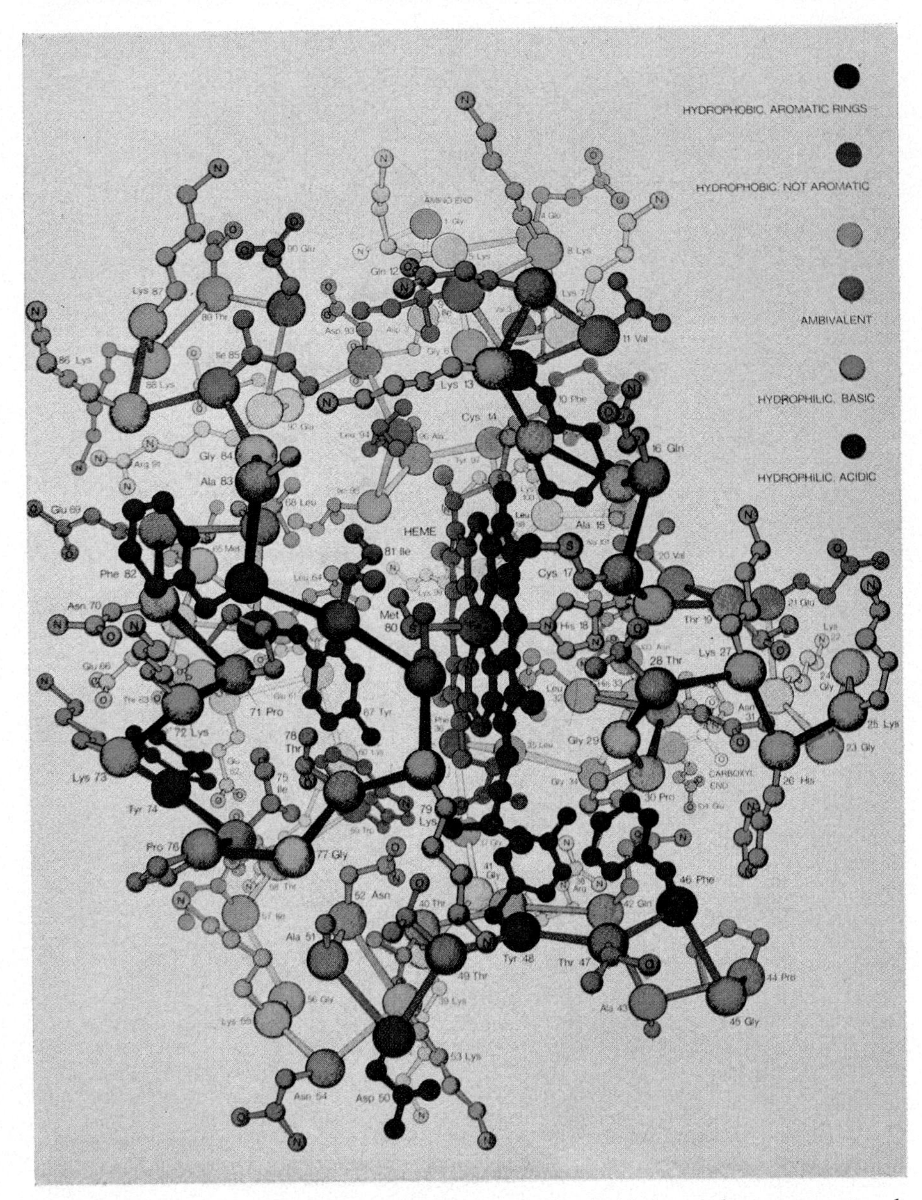

FIGURE 9 Structure of cytochrome c showing hydrophobic internal arrangement and hydrophilic external arrangement. [From *Scientific American*, April 1972. © 1972 by R. E. Dickerson and I. Geis.

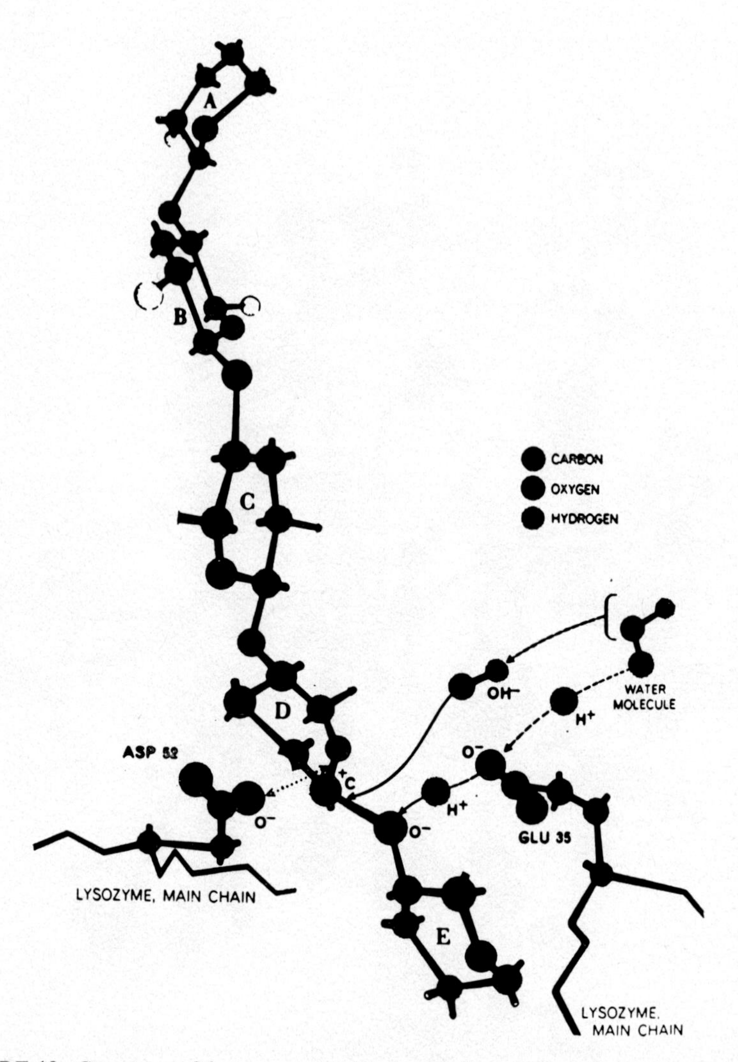

FIGURE 10 Structure of lysozyme showing mechanism of polysaccharide hydrolysis. [From D. C. Phillips, "The three-dimensional structure of an enzyme molecule." *Scientific American*, November 1966. © 1966 by Scientific American, Inc. All rights reserved.]

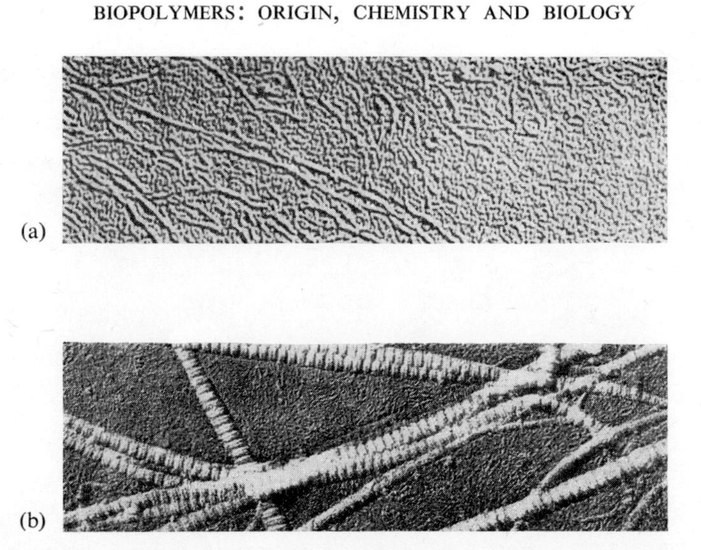

FIGURE 12 Structure of collagen: (a) Separated collagen fibrils, (b) Reaggregated fibrils.

alone. The protein is separated from the nucleic acid, redissolved and by changing the salt concentration so the material will reaggregate, with the results shown here. You can see that the material has reaggregated with the correct dimensions in one way, but not in the other; particles have all different lengths, in contrast to the particles of the native TMV virus; because there is no nucleic acid "information" to give the correct dimensions to the particle. In Figure 13c you can see the reconstituted TMV virus particle, which occurs when the nucleic acid component is put back into the protein solution. This same type of experiment has been done with much more complex particles, T4 bacteriophage, MS2 phage protein, etc. Progress in the reassembly of very complex structures is being made.[18] The TMV particle itself is shown diagrammatically in Figure 14. The tertiary structure of the subunit protein material in the TMV virus particle has not yet been established, but we know that there is a sequence of 120 amino acids and we know which ones they are, but, as yet, we do not know the detailed structure of this subunit.

INTERACTION OF PROTEIN AND LIPIDS

Protein–lipid interaction is one of the areas of very high scientific activity, in the technological polymer area and more selectively in the laboratory than biochemists have done in the past. Lipids have been used for handling biochemical proteins for some time. The present concept of membrane structure, showing the protein embedded in the bilipid membrane, is given in Figure 15.

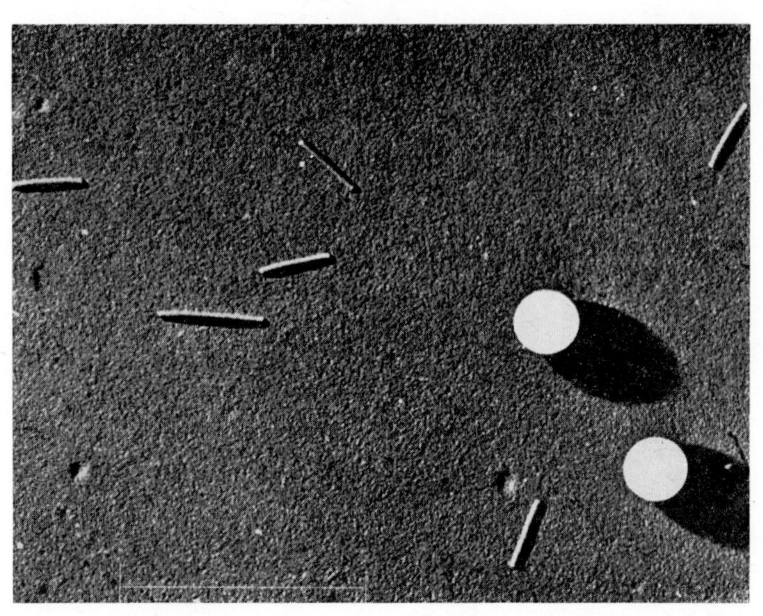

FIGURE 13a Tobacco mosaic virus (TMV), native.

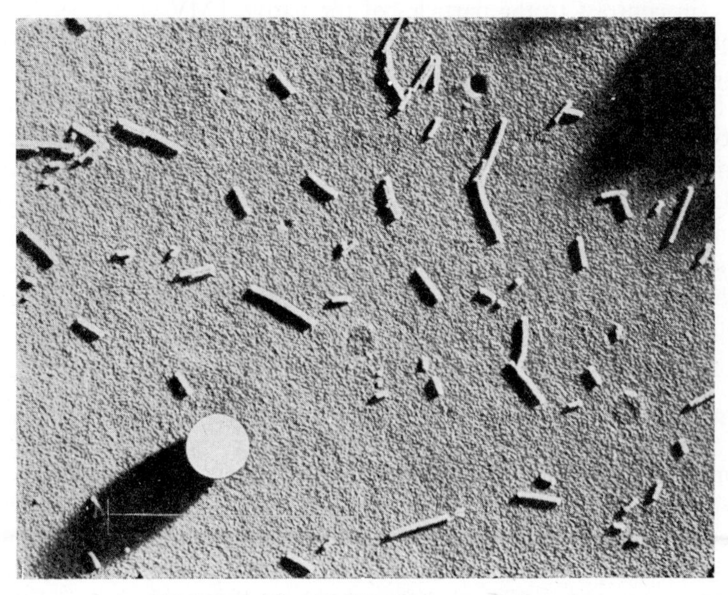

FIGURE 13b TMV protein, reconstituted.

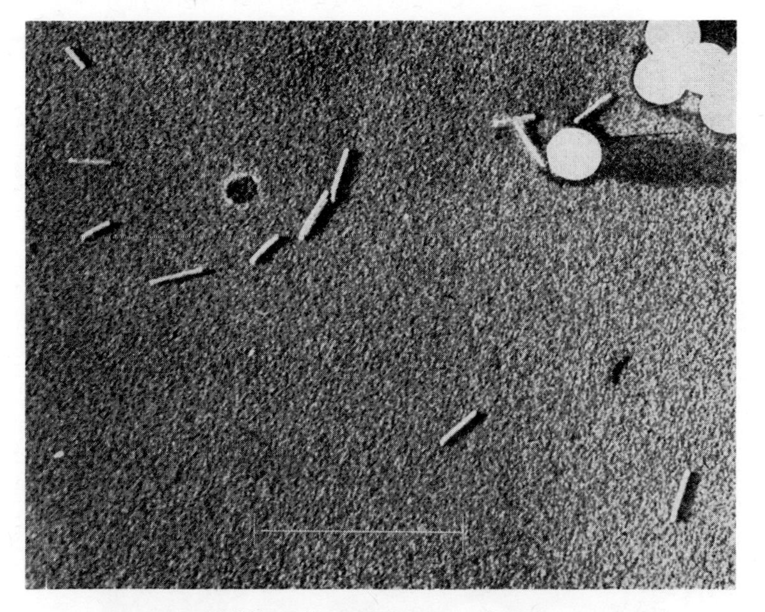

FIGURE 13c TMV, reconstituted.

The phospholipid consists of hydrophilic ends (the phosphoethanolamine end represented by spheres), and the lipid chains. The globules which are membrane proteins in various stages of insertion into the bilipid membrane make the total actual biological membrane itself. These proteins may be structural or they may be transport proteins, i.e., those carrying material into the cell through the biological membrane. The fact that many of the proteins are directly associated with lipids implies that there must be hydrophobic connections on the outside of the proteins in order to be able to perform this type of insertion. This has not yet been established in the laboratory, and I will give you more evidence for that guess later on. There are clearly hydrophobic parts on the proteins, but whether they are on the outside or the inside, or both, is a subject of some concern.

A synthetic "biological cell" is shown in Figure 16. It is made by shaking up a phospholipid (lecithin) with a protein cytochrome to form the liposomes (soft spheres). The multiple and single bilipid layers are visible in this electron micrograph, and these layers are filled with protein.[19] This work was done about eight years ago in a laboratory in Great Britain. I have called the spheres "Bangasomes" after the man who constructed them first.[20] The spheres are called generally liposomes, and their osmotic properties, transport properties, and a variety of other properties resemble those of living cells. How do materials get in and out through that membrane? What is the mode of their

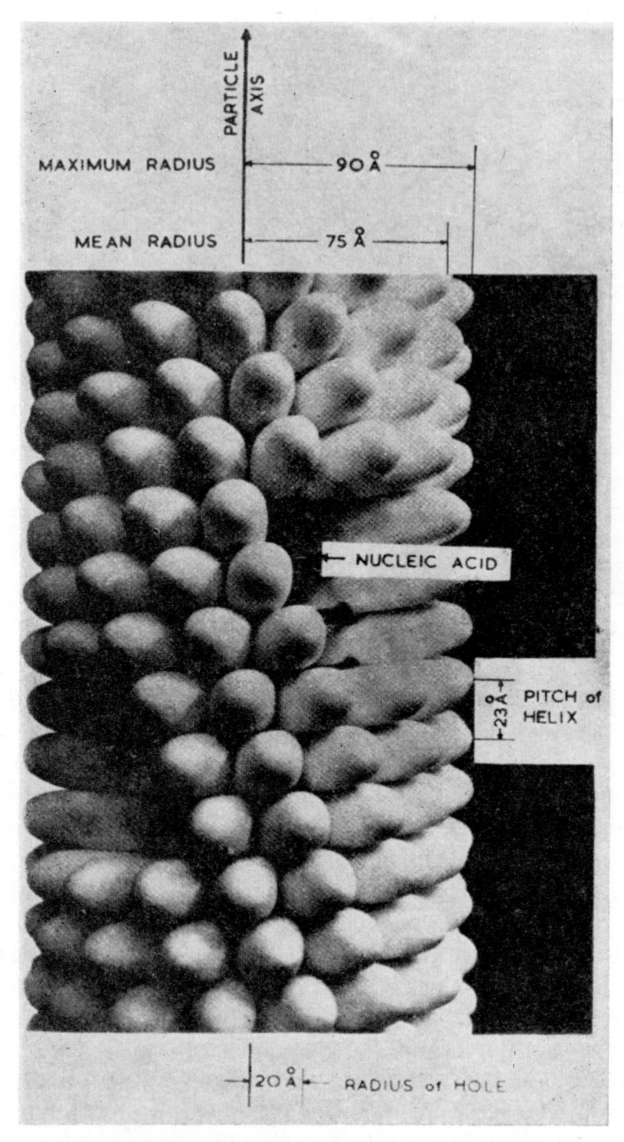

FIGURE 14 Diagrammatic structure of TMV.

transport? Can specific proteins be associated with the membranes to produce these specific results? These are questions still unanswered. Studies are now in progress, and this field is really just blossoming out as a new area and a new era. Only now have the polymer physical chemists got into the act,

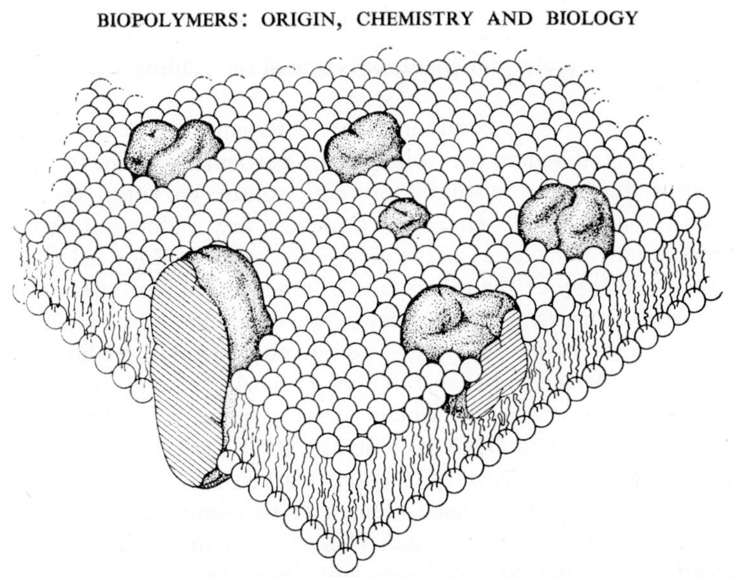

FIGURE 15 Present concept of membrane structure showing protein embedded in bilipid membrane.

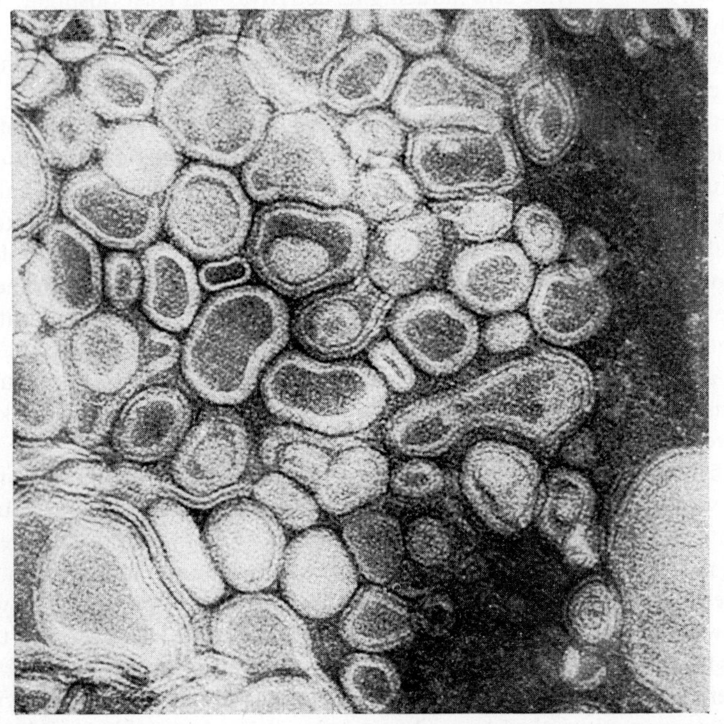

FIGURE 16 Liposomes of phospholipid with cytochrome (Horne and Watkins).

and Midland Macromolecular Institute is one of the leading sites of this type of research.

As you know, it is quite common to use detergents, anionic, cationic and non-ionic, to try and help in the isolation of a particular enzyme from a living cell. The procedures for the soluble proteins (soluble enzymes) are now well established. The insoluble enzymes, however, are a different story. The membrane-attached enzymes, as the membrane of the cell, or the membrane of the mitochondrion, or the membrane of other internal cellular structures, present more difficulties. It is necessary to use detergents of various types for enzyme extraction. A few years ago, we naïvely (and this quality has some merits as well as demerits) began this type of work. We had to learn that in order to extract the enzyme in which we were interested (which happened to be the RNA-instructed DNA polymerase, RIDP) detergent was required. I wondered why it was necessary. We also learned that when the detergents were eliminated, the enzyme itself seemed to be eliminated, or at least the activity of the enzyme was eliminated. This gave rise to the idea that perhaps the activity of the enzyme was dependent upon the presence of detergent molecules, i.e., the lipid component in some way was changing the conformation of the enzyme and inducing its activity, or at least raising it. Therefore, the residual enzyme activities which are observed when synthetic detergents were *not* used for extraction were simply due to the natural phospholipid which came out when the solution was sonicated.

We found that when we added small amounts of certain types of non-ionic detergents it was possible to get the enzyme activity back again. This constitutes a new development. No one really believed that the intrinsic activity of the enzyme was dependent upon its association with a detergent molecule. Up until now it has been considered that the detergent molecules were solubilizing entities to take the enzyme out of the lipid membrane. The detergent does, in fact, take the enzyme out of the lipid membrane, but large amounts of the detergent are not needed to make the enzyme active.

The activation of the RIDP enzyme by detergents is shown in Figure 17. RIDP is the enzyme which copies RNA into DNA, which is popularly known as "reverse transcriptase activity". Work on this enzyme is one of the major breakthroughs of the last two or three years, particularly in connection with tumor viruses. The activity of the enzyme is increased by addition of suitable amounts of various types of non-ionic detergents.[21,22] If you translate the effect into a molecular basis, this is a positive molecular interaction. This result occurred as a technical development of how to study the enzyme, but we did all the work on the detergents more or less as a side issue. However, this turned out to be the central theme, which is not uncommon in this type of work. We were actually studying the inhibition of the RIDP enzyme activity by certain drugs with a view toward using those drugs possibly to inhibit the transformation of normal cells into cancer cells by the viruses which carry this enzyme. The RIDP enzyme copies the RNA from DNA in the virus, which copy is then inserted into the DNA of the cell, thus making a transformed cell. The rationale was to find a drug which would inhibit the RIDP enzyme and thus block the transformation into cancer cells. In order to study the inhibition of the enzyme, we had to study inhibition of the enzyme activity under many conditions. This gave rise to our studies on detergent effect on enzyme activity. We found that we had some excellent inhibitors in the form of the drug itself. This particular drug, rifampicin and its derivatives is a lipophilic substance. Here, it turns out that if there is too much detergent the drug is ineffective in its inhibition of RIDP enzyme activity. The structure of the non-ionic detergents which we used in this study is shown in Figure 18. Some of the detergents (the Tritons) are aromatic and others are not. This fact is important. The effect of three different detergents on the ability of a certain drug, which has a hydrophobic tail and a hydrophilic end, to prevent the enzyme from working. The enzyme is shown in Figure 19. We have tried to plot the critical micelle concentration for each of the detergents, and you will note that the removal of the drug from its ability to inhibit the enzyme is dependent upon the formation of the micelles.[22] We interpret this to mean that if there is too much detergent present there is micelle formation, and the micelles dissolve the drug, taking it away from the hydrophobic portion of the enzyme. The drug no longer can act on the enzyme, so the enzyme returns to its full activity. In Figure 20 are two magnifications of a tissue culture which has been transformed into malignancy. You can see the growth of the cells to confluence here, but wherever the cells have been transformed into malignancy they overgrow each other, and they make little "piles" called foci. We have found that by suitable adjustment of the drug concentration it is possible to prevent the virus from transforming cells as determined by this focus formation inhibition.[23] We have gone even further with the drugs to show we can actually prevent the formation of tumors in whole animals.[24] Again, this requires a suitable combination of detergent and drug to prevent the detergent from spoiling the activity of the drug.

I hope that some of the things described above will give you some concept of the kind of a china shop you could get into, if you are willing to do it. I want to give you who are not in the biochemistry–molecular biology business some idea of the way in which a polymer chemist could effect the development of our fundamental concepts of the nature of life, and how it came about, and its application in the problems of the day.

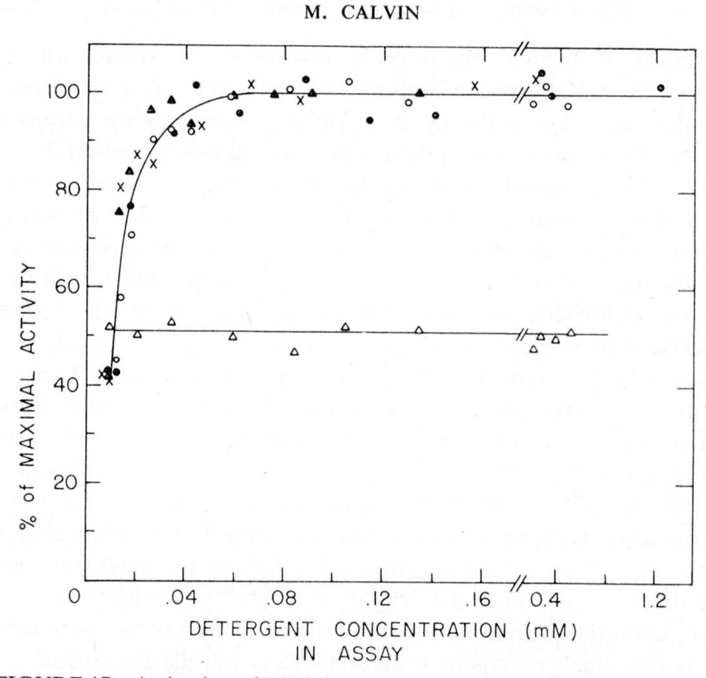

FIGURE 17 Activation of RNA-instructed DNA polymerase by detergent.

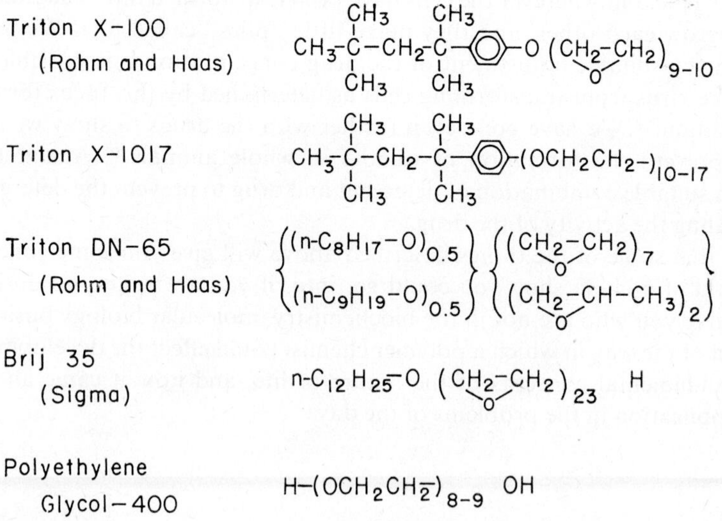

Triton X-100 (Rohm and Haas)	$CH_3-C(CH_3)_2-CH_2-C(CH_3)_2-C_6H_4-O\,(CH_2-\underset{O}{}-CH_2)_{9-10}\;H$
Triton X-1017	$CH_3-C(CH_3)_2-CH_2-C(CH_3)_2-C_6H_4-(OCH_2CH_2-)_{10-17}\;OH$
Triton DN-65 (Rohm and Haas)	$\left\{\begin{matrix}(n\text{-}C_8H_{17}-O)_{0.5}\\(n\text{-}C_9H_{19}-O)_{0.5}\end{matrix}\right\}\left\{\begin{matrix}(CH_2-\underset{O}{}-CH_2)_7\\(CH_2-\underset{O}{}-CH-CH_3)_2\end{matrix}\right\}H$
Brij 35 (Sigma)	$n\text{-}C_{12}H_{25}-O\,(CH_2-\underset{O}{}-CH_2)_{23}\;H$
Polyethylene Glycol-400	$H-(OCH_2CH_2-)_{8-9}\;OH$

FIGURE 18 Structure of detergents used in RNA-instructed DNA polymerase activation.

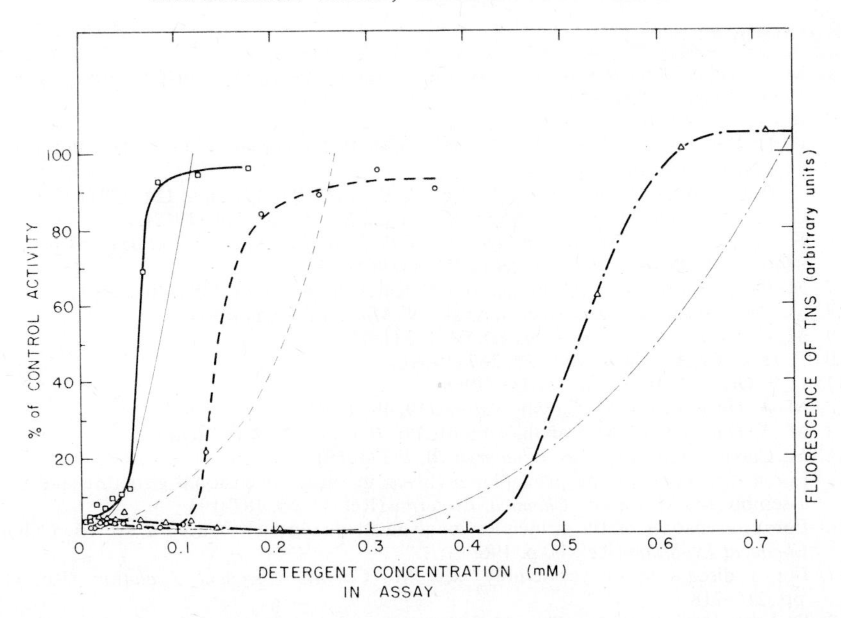

FIGURE 19 Suppression of drug effect on RNA-instructed DNA polymerase by detergents.

MSV Foci on BALB / 3T3 Culture

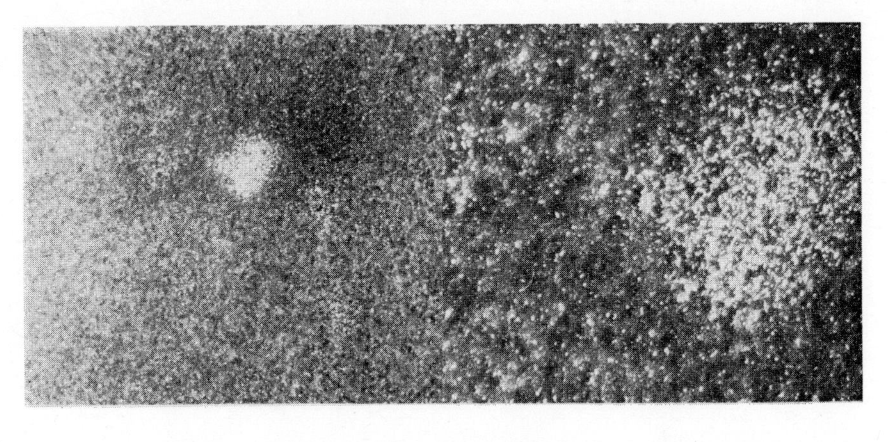

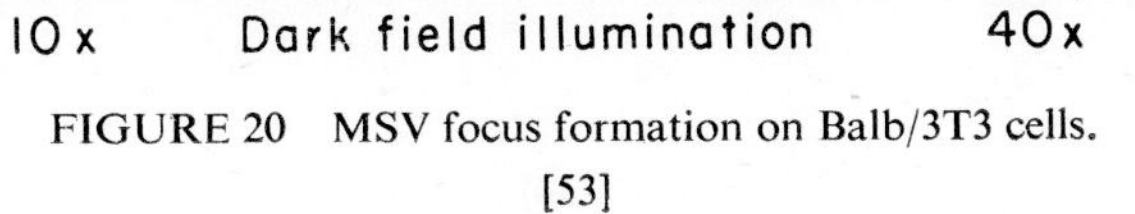

FIGURE 20 MSV focus formation on Balb/3T3 cells.

References

1. M. Calvin, *Chemical Evolution*, Oxford University Press, Oxford, England (and New York), 1969, pp. 152–155.
2. Ref. 1, pp. 161–169.
3. D. H. Kenyon and G. Steinman, *Biochemical Predestination*, McGraw-Hill Publishing Co., New York (1969), pp. 182–192.
4. D. H. Kenyon, G. Steinman and M. Calvin, *Biochim. Biophys. Acta* **124,** 339 (1966).
5. G. Steinman and M. N. Cole, *Proc. Nat. Acad. Sci. US* **58,** 735 (1967).
6. K. Harada and S. W. Fox in *Origins of Prebiological Systems and Their Molecular Matrices*, S. W. Fox, ed., Academic Press (1965), p. 296.
7. M. Paecht-Horowitz, J. D. Breger and A. Katchalsky, *Nature* **228,** 636 (1970).
8. M. Paecht-Horowitz and A. Katchalsky, *J. Molec. Evol.*, in press.
9. M. Eigen, *Die Naturwissenschaften* **58,** 465 (1971).
10. F. H. C. Crick, *J. Mol. Biol.* **38,** 267 (1968).
11. L. E. Orgel, *J. Mol. Biol.* **38,** 381 (1968).
12. M. A. Harpold and M. Calvin, *Nature* **219,** 486 (1968).
13. M. A. Harpold and M. Calvin, submitted to *Biochim. Biophys. Acta.*
14. M. Calvin, *Proc. Roy. Soc. Edinburgh* **70,** 273 (1969).
15. For a discussion of the principles involved in three-dimensional structure and self-assembly, see M. Calvin, *Chemical Evolution* (Ref. 1), pp. 187–193.
16. For a discussion of the principles involved in quaternary structure, see M. Calvin, *Chemical Evolution* (Ref. 1), p. 196.
17. For a discussion of reassembly, see M. Calvin, *Chemical Evolution* (Ref. 1), pp. 216–218
18. D. J. Kushner, *Bacteriol. Rev.*, **33,** 302 (1969).
19. D. Papahadjopoulos and N. Miller, *Biochim. Biophys. Acta* **135,** 624 (1967); D. Papahadjopoulos and W. Watkins, *Biochim. Biophys. Acta* **135,** 639 (1967).
20. A. D. Bangham, J. DeGier and G. D. Grevitch, *Chem. Phys. Lipids* **1,** 225 (1967).
21. F. M. Thompson, L. J. Libertini, U. R. Joss and M. Calvin, *Science* **178,** 505 (1972).
22. F. M. Thompson, L. J. Libertini, U. R. Joss and M. Calvin, submitted to *Biochemistry.*
23. M. Calvin, *Radiation Res.* **50,** 105 (1972).
24. U. R. Joss, A. M. Hughes and M. Calvin, submitted to *Nature.*

Polymers in Medicine†

DONALD J. LYMAN

Division of Artificial Organs (College of Medicine) and Division of Materials Science and Engineering (College of Engineering), University of Utah, Salt Lake City, Utah

The use of polymers in repair of the vascular system and in extracorporeal handling of blood is reviewed. Discussed are the adsorption of protein on surfaces as studied *via* infrared multiple internal reflectance spectroscopy, *in vivo* experiments involving implanted cannulas, pre-coating of polymers with albumin, and the adhesion of baby hamster kidney cells to polymer surfaces.

The use of materials in the body to repair or restore damaged, diseased, or ravaged tissue and organs is not new. First recorded use was that of a gold plate in 1588 to repair a cleft palate. Later in the 1800's, there were numerous reports of metal plates and pins to fix broken bones. However, with the advent of the polymer industry and the ready availability of a variety of materials (polymers) having properties more similar to the body, there was a tremendous increase in the use of materials in surgery. Most of this increase being since the middle 1950's. These uses have been quite broad, ranging from temporary assist materials, such as sutures, staples, surgical adhesives, plasma extenders, bone pins and braces, to relatively simple artificial parts of a more permanent nature, such as vascular grafts, heart valves, hydrocephalic drain tubes, joints, reinforcing meshes, as well as a variety of soft tissue replacement materials for cosmetic surgery, to the more complex devices such as the artificial kidney, the artificial lung, and the artificial heart, which can duplicate some physiological process. Indeed the imagination and skill of the surgeon and the support and technical assistance from companies such as Midland's own Dow Corning Center for Aid to Medical Research have resulted in a great variety of devices being available (See Figure 1). However, even with all these successes there have been many failures. Most of this has been due to improper choice of material for the intended use. These materials are not

†Lecture presented at the Scientific Symposium "Trends in Macromolecular Science", at the Dedication Ceremony of the Midland Macromolecular Institute, September 29, 1972.

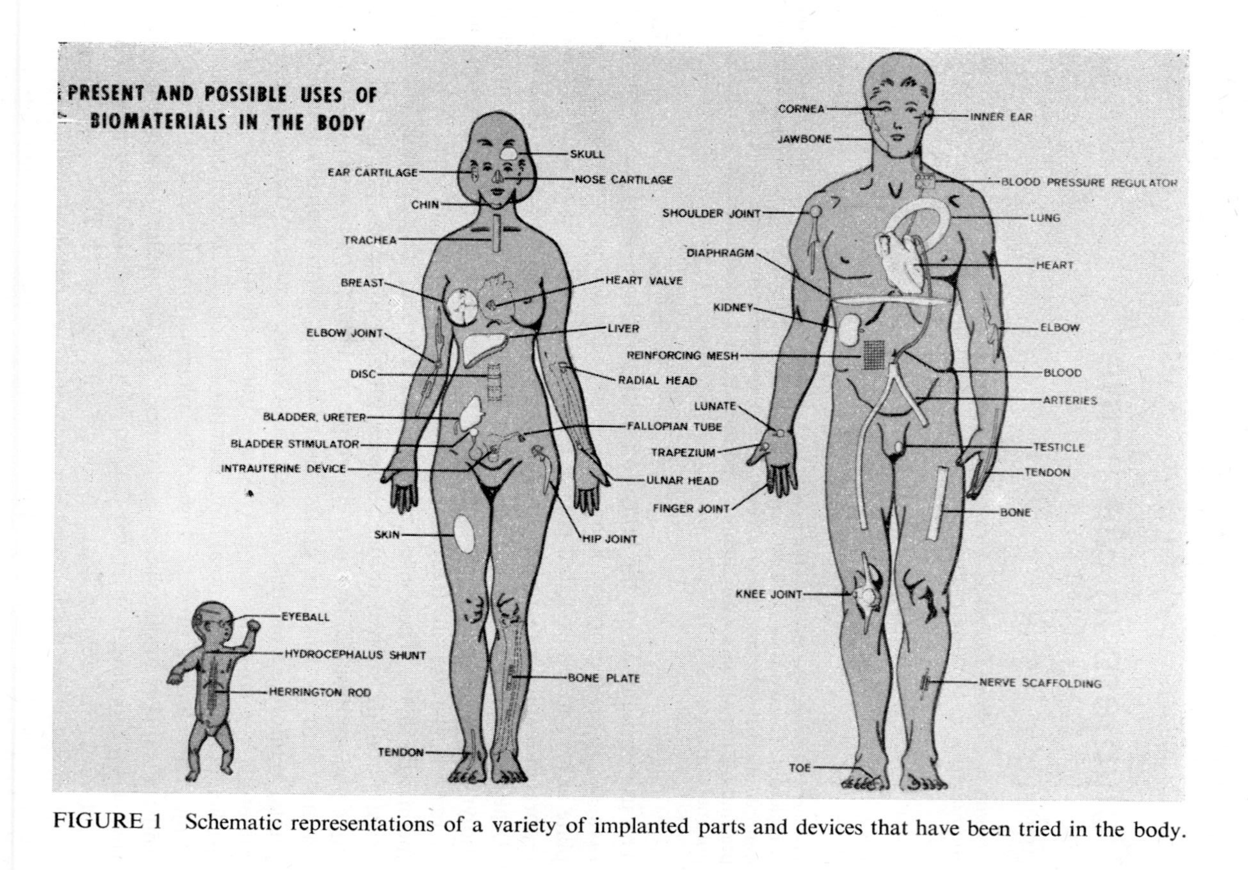

FIGURE 1 Schematic representations of a variety of implanted parts and devices that have been tried in the body.

inert in the body, but do react on and are in turn acted on by the biological environment.[1] Also, the materials often tried were those readily available from many commercial processes. These do not have the purity and reproducibility that we would like in a biomedical material.

As a result, we often find that medical advances have been highly dependent on new polymers becoming available from a variety of industrial or consumer needs, which might also meet these medical needs. During the 1960's, a new research area has developed with the objectives of providing the knowledge of how polymers interact as they interface with the living system so that needed biomedical polymers can be developed in their own right. This research requires the coupling of many traditional disciplines as well as creating new interdisciplinary studies (see Figure 2).

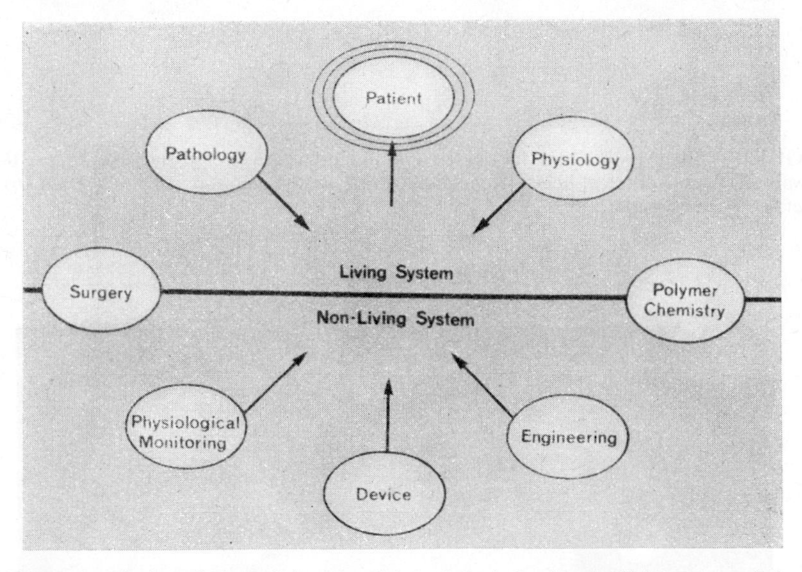

FIGURE 2 The interfacing of the material (and device) with the living system is key to determining the fate of the patient. To study these interface reactions involves the coupling of many disciplines.

The area that I have found most fascinating is the use of polymers in repair of the vascular system and in extracorporeal handling of blood. I would like to describe this work to you, using the artificial heart as the background for my remarks.

For some time, the Division of Artificial Organs, under the direction of Dr. W. J. Kolff, have been developing artificial heart devices. Much of this work has been concerned with a Kwan-Gett hemispherical heart[2] (see Figures 3 and 4) and the development of optimum surgical procedures as well as pre- and post-

FIGURE 3 The molds and fabricated copolyurethane housing and diaphragm for a Kwan-Gett hemispherical heart. [Reprinted from *Trans. Amer. Soc. Artific. Internal Organs* (Ref. 3), by permission.]

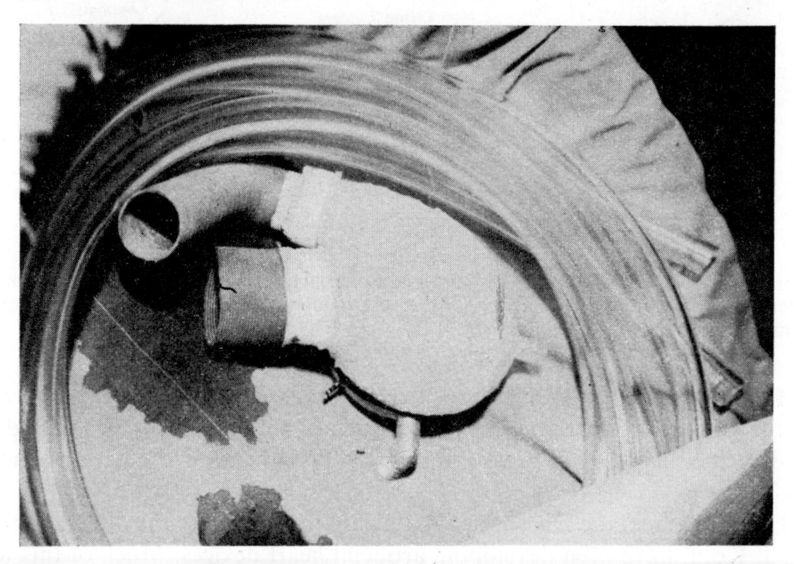

FIGURE 4 An assembled Kwan-Gett hemispherical heart (right ventricle) just prior to implantation. [Reprinted from *Trans. Amer. Soc. Artific. Internal Organs* (Ref. 3), by permission.]

operative care techniques. The Division has made considerable progress as indicated by a world's record survival time of two weeks in a calf. However, there are still problems; such as occasional mechanical failure, poor fit, inadequate pumping, improper blood pumping pressure, need for improved valves, etc. Also, we still do not have an ideal material to interface with the blood. As a result we find clots on the surface of the artificial heart, and evidence of clots which broke loose forming emboli and have damaged distant tissues (see Figure 5). The cellular elements such as red blood cells and platelets are damaged and destroyed (see Figure 6). The protein constituents are also damaged (see Figure 7). While these studies do give an overall picture of the effect of the material on blood, there are two factors that must be kept in mind. First, the animals own defense mechanisms can often clear the damage, thus masking a potential effect. Second, these tests, in themselves, give no clues as to the initiating steps in this damage. We developed[4] a simple *ex-vivo* test which has allowed us to look more closely at this problem. For example, blood flowing over a polyvinyl chloride surface for several minutes in the absence of an air interface, a clot does form (see Figure 8). One can see a platelet on the surface, with fibrin strands extending from the platelet pseudopods, and this fibrous mesh entrapping red blood cells. However, while this has been valuable for guiding research and for screening polymers,[4,5] it still does not explain why the platelets adhere, and to what surface they are actually adhering. Our work has been directed toward answering these questions as well as the related one of how does one synthesize a non-thrombogenic polymer. This requires developing knowledge of the events occurring at the molecular level as the blood comes in contact with the polymer which initiate these adverse reactions. To do this, one must use model experiments (*in-vitro* and *ex-vivo*) to isolate and study the individual events that *might* occur at the molecular level in the *in-vivo* situation.

Blood coagulation has been studied for over 100 years, with many studies being directed toward identification of the activating molecule. These studies reached a high point in 1964 with the proposal of an enzymatic cascade blood clotting mechanism[6] resulting from surface contact of a particular protein molecule.

Our studies began here; on how proteins are adsorbed and activated on polymer surfaces. One tool that we have used extensively in this work has been infrared multiple internal reflectance spectroscopy. This allowed us to examine directly the adsorbed protein–polymer substrate complex, and not rely on indirect methods to estimate the interaction. Since both the adsorbed protein and the polymer infrared adsorption do appear in the spectrum (see Figure 9) one can use ratios of bands to quantitate how much protein is adsorbed[7] and the rates of adsorption.[8] We studied the adsorption of the more common blood proteins on a variety of neutral, hydrophobic polymers.

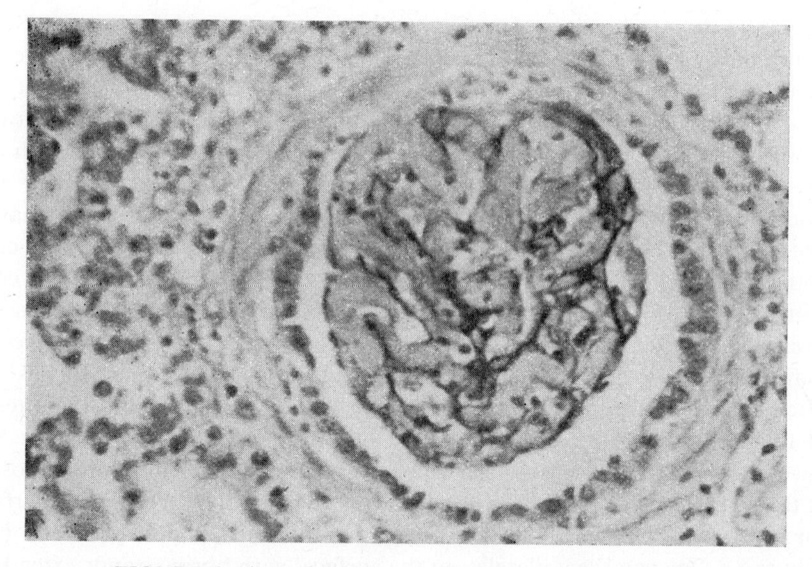

FIGURE 5 An embolus in a small blood vessel in the brain.

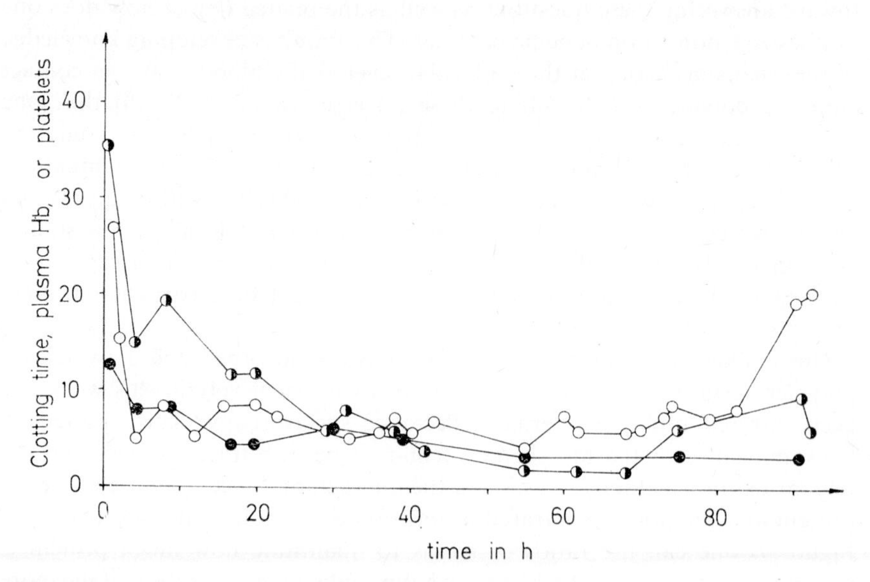

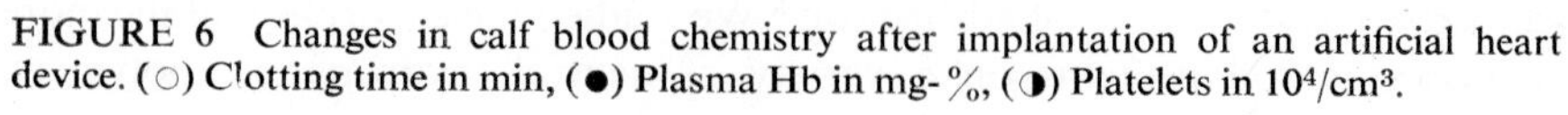

FIGURE 6 Changes in calf blood chemistry after implantation of an artificial heart device. (○) Clotting time in min, (●) Plasma Hb in mg-%, (◑) Platelets in $10^4/cm^3$.

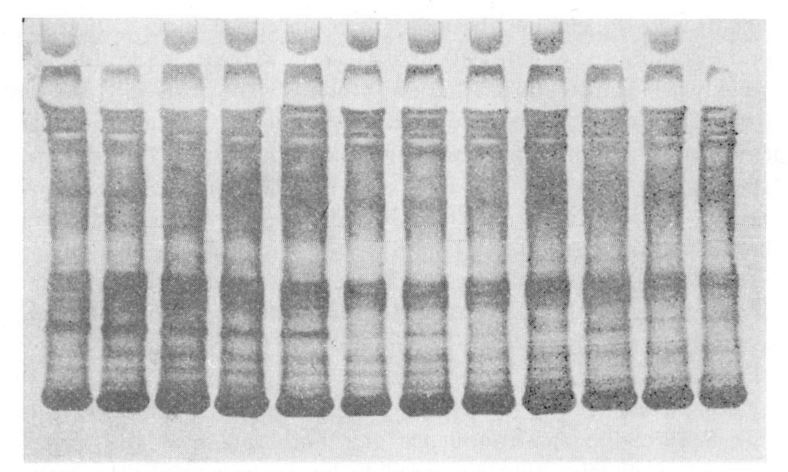

FIGURE 7 Changes in the electrophoretic patterns of plasma with time from a calf with an artificial heart.

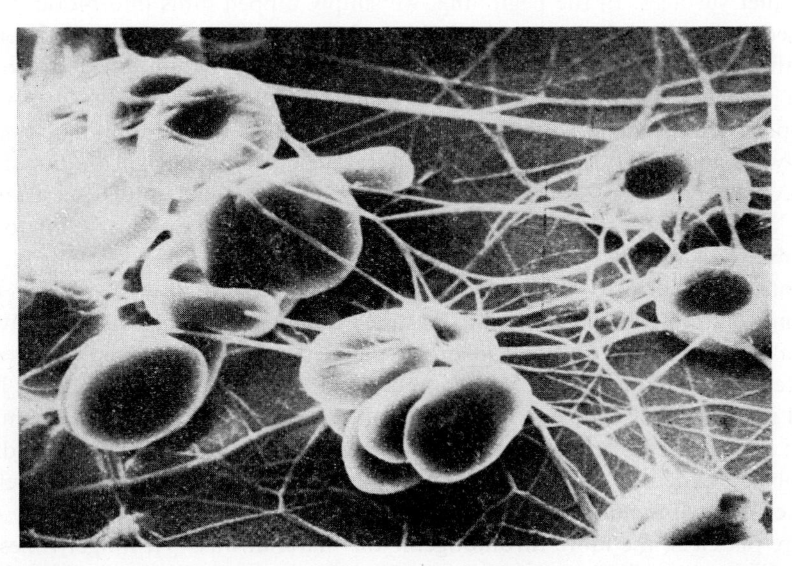

FIGURE 8 Scanning electron microscope picture of a clot formed on a polyvinyl chloride surface in a flow-through *ex-vivo* cell showing platelets on the surface of the material and the fibrin mesh entrapping red blood cells. [Picture by Dr. Rodman, University of Iowa.]

Experiments were done in the absence of an air interface under static and flow adsorption conditions. Figures 10 and 11 show typical adsorption curves. Very quickly it began to appear that the proteins were physically and essentially irreversibly adsorbed to these neutral hydrophobic polymers in a similar manner forming a monolayer. Electron micrographs and infrared spectra data (showing no shifts in Amide I—Amide II bands) indicated that no gross structural alterations occurred, i.e. the adsorbed proteins retained their globular form. As these studies continued we did begin to find differences in rates of adsorption, plateau concentration, effect of flow velocity in amount of adsorbed protein, etc., as we changed the adsorbing protein on the polymer surface. I will discuss this in more detail later.

At this stage, the *in vivo* experiments involving implanted cannulas that we were doing concurrently with the help of Dr. William Edmark (Providence Hospital, Seattle), began to influence our thinking. On all of the surfaces we examined, a platelet plug appeared to be at the site of clot initiation. Indeed, with several surfaces, the total clot appeared to be an aggregated platelet mass. As I pondered these contrasting sets of experiments, I began to feel that there was an overemphasis on "contact surface activation" of Hageman factor (Factor XII) and that the mechanism on the implanted polymer surface might more nearly resemble what happens when the natural vessel wall is cut. We began to study the adhesion of platelets to various polymer surfaces. In the beginning, we simply dipped films into blood, then rinsed and stained the surfaces. All the surfaces had large numbers of platelets on them, and all appeared surprisingly similar. However, since it was known that proteins denature at an air interface and it was possible that we were simply coating our polymer with denatured protein by the dipping—essentially a Langmuir transfer of protein to the surface, we redesigned our system to avoid an air interface. The flow-through cell was quite simple (see Figure 12), a small glass chamber to which we could clamp the polymer films to be examined. The cell system is filled with buffered saline. Human venous blood is introduced directly (via a vena puncture) to the cell, thus displacing the filling solution. Much care had to be taken to avoid air bubbles. The cell was then quickly rinsed and the adsorbed platelets fixed and stained. On some surfaces, you could actually see pseudopod development (see Figure 13), and if the time of exposure was longer, an actual clot which would resemble that shown earlier in Figure 8. As we examined a series of neutral hydrophobic polymers under these conditions and compared the number of platelets adhering at one-minute exposure, there was a definite trend observed: the more platelets, the more thrombogenic the material was.[4] At longer times of exposure, all surfaces began to look rather similar. This is not unexpected, since all were considered thrombogenic, with the only differences being in how long they took to clot.[3,9,10]

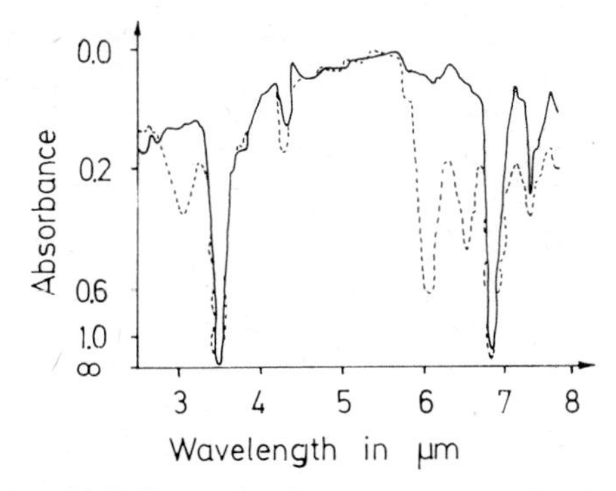

FIGURE 9 Infrared multiple internal reflectance spectra of polyethylene (———) and of γ-globulin adsorbed on polyethylene (– – – – –).

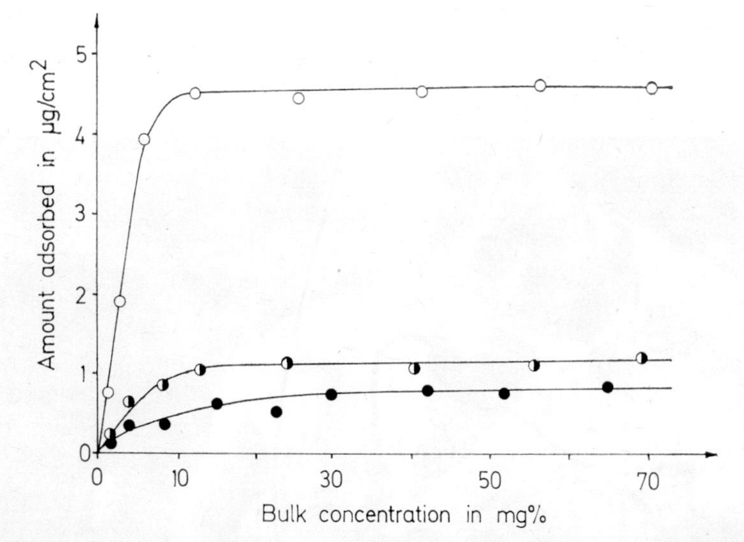

FIGURE 10 Adsorption isotherms of albumin on selected polymer surfaces. (○) Co (polyether/urethane/urea), PEUU-1, (◑) silastic rubber, (●) fluorinated poly(ethylene-co-propylene).

Since our protein adsorption studies indicated that a polymer surface could be coated with a layer of protein without altering the protein structure, we explored the possibility of passifying a polymer by pre-coating it with albumin. Indeed, when we tested such pre-albuminated surfaces in our flow-through platelet test cell, we found essentially no platelet adhesion.[11] If our hypothesis was true, this albuminated surface should be non-thrombogenic. Dr. Lande using albuminated surfaces in his membrane artificial lung[12] also showed markedly reduced platelet damage. Later, Dr. Andrade prepared

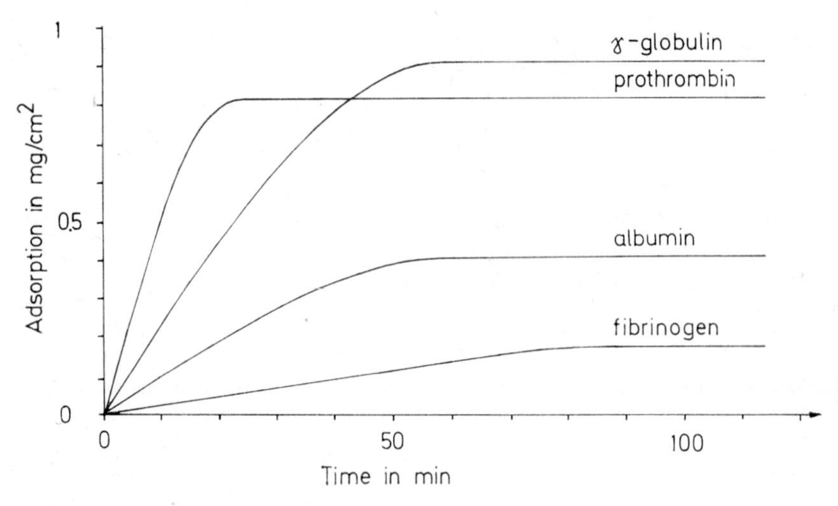

FIGURE 11 Adsorption of various proteins to fluorinated poly(ethylene-co-propylene). [Reprinted from *Advances in Nephrology* (Ref. 22), by permission]

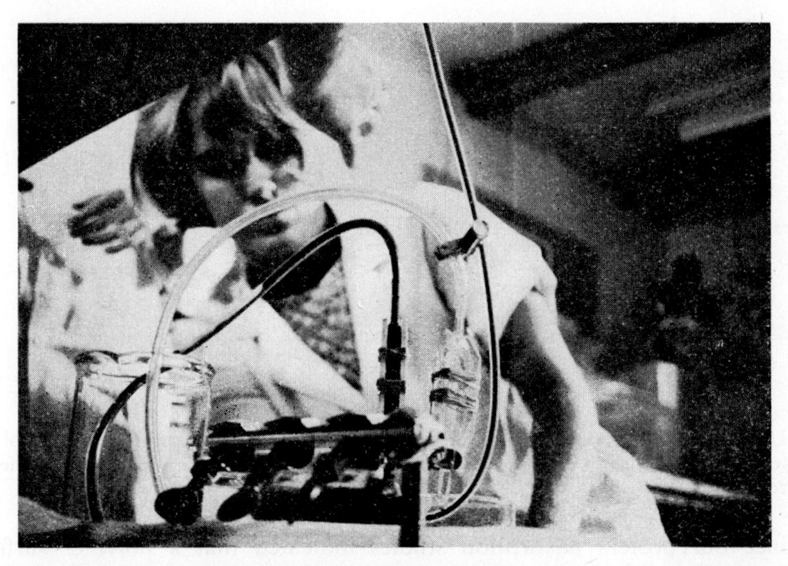

FIGURE 12 Two *ex-vivo* flow-through platelet adhesion cells; the left cell with venous blood flowing through it; the right cell primed with buffered saline.

covalently bonded albumin surfaces on Gott vena cava rings[13] which were then shown to be as non-thrombogenic as heparinized surfaces.[5] Thus, if we could avoid platelets adhering to a surface we should be able to avoid the formation of a clot.

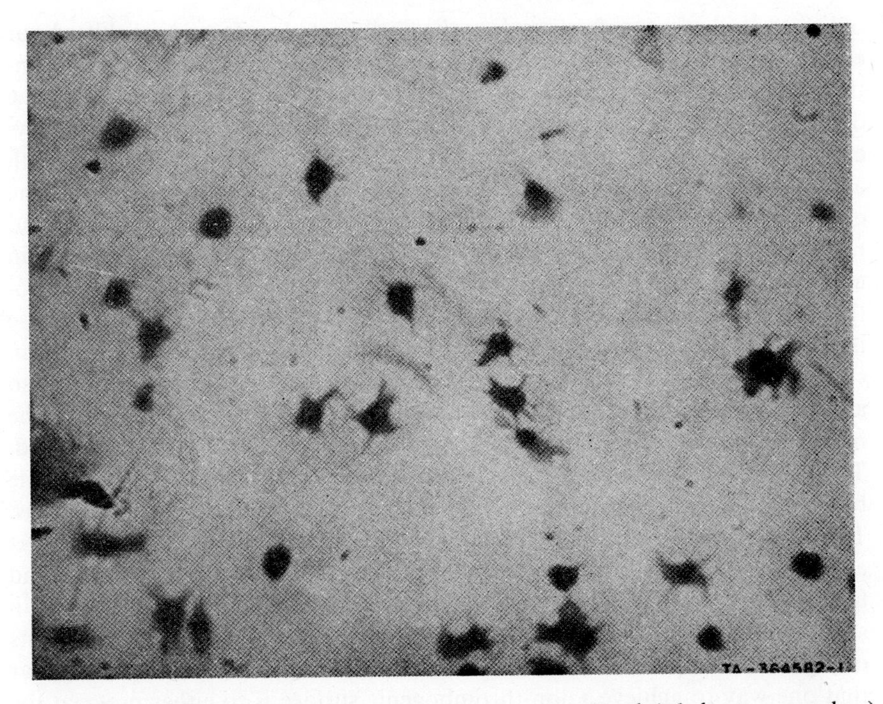

FIGURE 13 Stained platelets adhering to a fluorinated poly(ethylene-co-propylene) surface after exposure in our flow-through *ex-vivo* cell. [Reprinted from *Thrombosis et Diathesis Haemorphagic* (Ref. 4), by permission.]

With the interest in materials for artificial heart devices, we begin to explore a variety of segmented copolyether–urethanes and urethane–ureas for possible candidates. This family of polymers can be tailored to give a great variety of mechanical properties. One of these, a copolymer based on polypropylene glycol, methylene bis(4-phenylisocyanate) and ethylene diamine appeared to have a fair balance of mechanical and fabricating properties. Examination of this copolyether–urethane–urea in our platelet cell gave surprising results: i.e. it showed very low platelet adhesion.[15] Also, when Gott rings were implanted in the inferior vena cava of dogs, they were rated as non-thrombogenic as the heparinized surfaces.[14]

We did some preliminary studies on implanting a polyurethane hemi-spherical Kwan-Gett heart.[3] Although we did not get long survival in these initial implantations, the heart, on removal, was as clean and shiny as when we put it in. There was no evidence of clots, fibrin deposition or platelet deposits anywhere in the heart. Because of the heavy scheduling of other types of hearts, and other surfaces such as the fibril coated surfaces,† we shifted our efforts back to gaining an understanding of how these two types of non-thrombogenic surfaces work. What were the similarities, what were the differences between the albumin coated surface and the segmented copoly-ether–urethane–urea surface. Since precoating a polymer surface with albumin pacifies the surface and prevents platelet adhesion, can this same effect be achieved *in situ*, i.e. could the urethane surface preferentially and rapidly adsorb albumin from the blood? As we examined the nature of protein adsorption to polymer surfaces in more detail,[8] data supporting this hypothesis began to emerge. In the series of polymer surfaces examined, those having higher rates of albumin adsorption and also having higher plateau concentrations of albumin on the surface were less thrombogenic. There also appeared to be an inverse relationship for γ-globulin adsorption. That is, the more γ-globulin on the surface, the more thrombogenic the polymer appeared. This is, of course, assuming that adsorptions occurring in a mixture, i.e. whole blood, would be similar to that observed for simple, single solute adsorptions. This is currently being checked. However, supporting this is our platelet adhesion studies which shows that both γ-globulin coated surfaces and fibrinogen coated surfaces are more active toward platelets[8] than are albumin surfaces, and work by other investigators[17,18] showing the adverse effects of γ-globulin on platelet adhesion and release reactions. Thus it would appear that one way to achieve a non-thrombogenic surface is to either pre-coat the polymer with a pacifying layer of albumin or have a polymer of such a chemical structure that it achieves this pacifying albumin coating rapidly *in situ*.

Our work is currently proceeding along several lines of investigation: (1) defining the nature of the protein–polymer interaction; (2) determining the actual chemical and inherent water structure of the proteinated surfaces; and (3) determining the nature of the platelet–protein surface interaction. These studies, I feel, will allow us to develop much of the basic knowledge of blood interactions with polymeric surfaces; knowledge which is necessary if we are to advance the repair of the vascular system and the extracorporeal treatment of blood.

Recently, there has been increased interest in cell culturing on surface as an alternate way to pacify surfaces.[19] The neo-intima which develops provides a natural surface to the blood. However, lack of cellular adhesion to the Silastic Rubber substrate has necessitated the use of Dacron fiber-coated surfaces to prevent the cellular deposits from being stripped off. However, this is not without its own problems. Therefore, Dr. Hill and I investigated the adhesion of baby hamster kidney (BHK-21) cells[20] to polymer

†It should be noted that after much implantation and post-operative study by the Division of Artificial Organs, the world's record long-term survival was achieved with a fiber-coated surface.[16]

surfaces. Normally, if the BHK-21 cell adheres, it will grow to form a monolayer; if it does not adhere, it will ball up and die (see Figure 14). As we examined a series of copolyether-urethane-ureas, which were chemically identical except for the length of the polyether segment, we noticed an interesting phenomena.[21] When the polyether segment had a molecular weight of 400, 700 or 2000, the BHK-21 cells adhered and grew normally; when the polyether segment had a molecular weight of 1000, the cells did not adhere and grow. Parallel experiments on platelet adhesion show similar response; i.e. platelet adhesion on the polymer having the 400, 700, or 2000 molecular weight polyether, but not the 1000 molecular weight polyether. These four types of copolymers showed similar critical surface tensions and protein

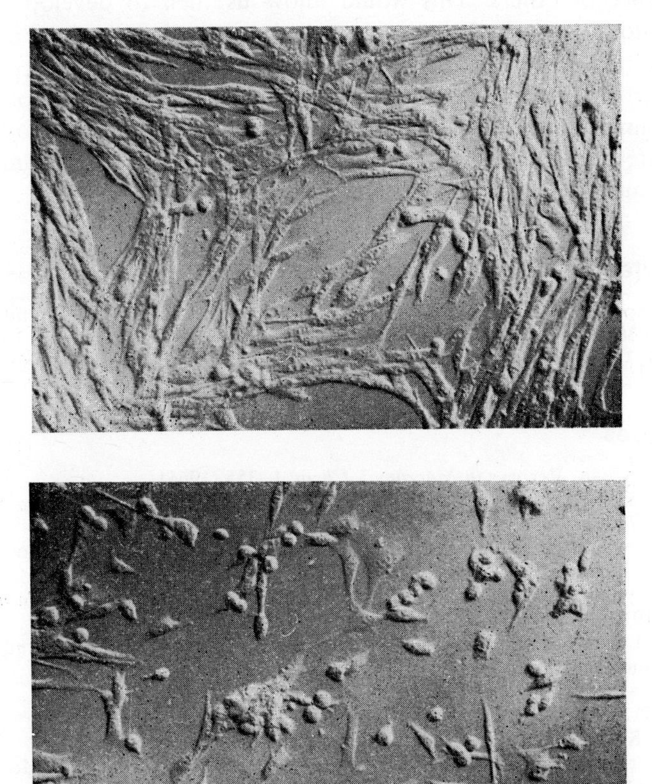

FIGURE 14 Cell culturing of baby hampster kidney cells: (a) onto glass, showing adhesion and growth, (b) onto a polyurethane surface, showing no growth.

adsorption. However, it must be kept in mind that these involve gross measurements, and the cells are apparently recognizing some microarchitectural differences in these copolymers. Indeed, preliminary studies on the tacticity of the polypropylene glycol segments and on the film morphology does indicate that the copolyether-urethane-urea based on the 1000 molecular weight glycol is different from the others. We are now working to define the structure of these surfaces and couple this information with our other blood studies. These studies have double interest to us. Platelets are rather difficult to handle, and so if we do have a cell line that mimics their action, we could do more quantitative studies on the nature of the polymer surface interaction. Also, if each cell line has its own surface recognition pattern, we might be able to make a surface that will preferentially adhere (and grow) one cell line in the presence of others. This would allow us then to develop scaffolding polymers to assist the body in the controlled repair of its own tissue and organs; whether it be a nerve, the trachea, or a blood vessel.

To me, the study of polymers in the body is a most exciting one, and the future is unlimited since the knowledge being developed can also be applied toward preventive medicine. I want to thank the Dedication Committee for allowing me to tell you about some of our work in this area.

Acknowledgments

The work reported here was supported in part under a National Science Foundation Grant No. GK-29382, a National Institute of Arthritic and Metabolic Disease Contract No. NIH-70-2017, and the University of Utah Biomaterials Research and Development Fund.

References

1. D. J. Lyman, *Reviews in Macromol. Chem.* **1,** 355 (1966).
2. C. Kwan-Gett, H. H. H. J. Zwart, A. C. Kralios, T. Kessler, K. Backman and W. J. Kolff, *Trans. Amer. Soc. Artif. Int. Organs* **16,** 409 (1970).
3. D. J. Lyman, C. Kwan-Gett, H. H. H. J. Zwart, A. Bland, N. Eastwood, J. Kawai and W. J. Kolff, *Trans. Amer. Soc. Artif. Int. Organs.* **17,** 456 (1971).
4. D. J. Lyman, K. G. Klein, J. L. Brash and B. K. Fritzinger, *Thromb. Diath. Haem.* **23,** 120 (1970).
5. D. J. Lyman, K. G. Klein, J. L. Brash, B. K. Fritzinger, J. D. Andrade and F. S. Bonamo, *Thromb. Diath. Haem.* (Suppl. **42**), 109 (1970).
6. R. G. MacFarlane, *Nature* (London) **202,** 498 (1964).
7. J. L. Brash and D. J. Lyman, *J. Biomed. Mater. Res.* **3,** 175 (1969).
8. S. W. Kim, R. G. Lee and D. J. Lyman, Manuscript in preparation.
9. L. I. Friedman, H. Lien, E. F. Grabowski, E. F. Leonard and C. W. McCord, *Trans. Amer. Soc. Artif. Int. Organs* **16,** 63 (1970).
10. D. J. Lyman, *Trans. Amer. Soc. Artif. Int. Organs* **16,** 75 (1970).
11. *Chem. & Eng. News*, January 27, 1969—Page 37.
12. A. J. Lande, L. Edwards, J. H. Bloch, R. G. Carlson, V. Subramanian, R. S. Ascheim, S. Scheidt, S. Filmore, T. Killip and C. W. Lillehei, *Trans. Amer. Soc. Artif. Int. Organs* **16,** 352 (1970).
13. V. L. Gott and A. Furuse, *Fed. Proc.* **30,** 1679 (1971).

14. V. L. Gott, M. D. Ramor, F. B. Najjar, J. C. Allen and K. E. Becker, *Proc. Artificial Heart Program Conference*, edit. by R. J. Heggeli, Washington, D.C., U.S. Govt. Printing Office, 1969, Chap. 17, p. 151.
15. D. J. Lyman, J. L. Brash and K. G. Klein, *Proc. Artificial Heart Program Conference*, edit. by R. J. Heggeli, Washington, D.C., U.S. Govt. Printing Office, 1969, Chap. 11, p. 113.
16. E. J. Hershgold, C. J. Kwan-Gett, J. Kawai and K. Rowley, *Trans. Amer. Soc. Artif. Int. Organs* **18,** 181 (1972).
17. M. F. Glynn, M. A. Packham, J. Hirsh and J. F. Mustard, *J. Clin. Invest.* **45,** 1013 (1966).
18. J. F. Mustard, M. F. Glynn, E. E. Nishizawa and M. A. Packham, *Fed. Proc.* **26,** 106 (1967).
19. C. G. LaFarge and W. F. Bernhard, *Proc. Artificial Heart Program Conference*, edit. by R. J. Heggeli, Washington, D.C., U.S. Govt. Printing Office, 1969, Chap. 52, p. 619.
20. J. C. Taylor, D. W. Hill and M. Rogalsky, *Exp. Cell Res.*, in press (1972).
21. D. J. Lyman, D. W. Hill, R. K. Stirk, C. Adamson and B. R. Mooney, *Trans. Amer. Soc. Artif. Int. Organs* **18,** 19 (1972).
22. D. J. Lyman and S. W. Kim, *Advances in Nephrology* (from the Necken Hospital) **2,** 96 (1972) J. Hamburger, J, Crosnier, M. H. Maxwell, eds. Year Book Medical Publishers, Inc. Chicago.

Molecular Configuration in Bulk Polymers†

PAUL J. FLORY

Department of Chemistry, Stanford University, Stanford, California 94305

Experimental results on the following topics are examined and discussed from the point of view of their bearing on the molecular configurations and intermolecular correlations in amorphous polymers: (i) the effect of dilution on the force of retraction f in stretched elastomers; (ii) the effect of dilution on the force-temperature coefficient, and the correspondence of $-[\partial \ln(f/T)/\partial T]_{V,L}$ to $d\ln < r^2 >_0/dT$ found for the linear polymer in dilute solution; (iii) comparison of experimental cyclization constants K_x for siloxanes, both in absence and in presence of an inert diluent, with values calculated from dimensions ($< r^2 >_0$) for the linear polymer; (iv) thermodynamic activities of solutions in the Henry's law range; (v) meager results currently available from direct determination of dimensions of polymer chains in the bulk polymer; (vi) depolarized light scattering and the effect thereon of dilution with an isotropic diluent; and (vii) strain birefringence and the effects of dilution on the stress-optical coefficient. Optical anisotropies from (vi) and (vii) and their dependence on concentration indicate local intermolecular correlations, which, however, appear to be not much greater than for simple liquids. None of the experiments (i) to (iv) gives any intimation of an effect of dilution that could be ascribed to dispersal of an ordered arrangement of chains. Results from (ii) and (iii) demonstrate that the same chain configurational parameters found in dilute solutions hold quantitatively in the bulk polymer. Evidence is thus compelling that chain configurations in the bulk amorphous polymer differ inappreciably from the configurations in a dilute solution, apart from effects of excluded volume in the latter environment.

---

†Lecture at the Scientific Symposium at the occasion of the Dedication of Midland Macromolecular Institute, September 29, 1972. The paper will be published in full in *Pure and Applied Chemistry*.

Structure–Property Relationships in a Polymer†

E. H. ANDREWS

Department of Materials, Queen Mary College, London E1, England

The way in which polymer molecular structure controls structure on higher levels of organization in a solid polymer is briefly reviewed, as is also the way in which structure, on all levels, controls physical properties. The "line of descent" from molecular structure to physical properties is then illustrated at length in the case of one particular polymer, *cis*-polyisoprene (natural rubber). It is shown how the crystalline-amorphous morphology in the solid is controlled both by the chemical microstructure of the polymer and by the physical conditions (temperature, time, strain) under which solidification occurs. By changing these "processing conditions" great changes can be effected in the morphology.

The mechanical properties of the solid are then examined as a function of morphology and shown to depend strongly on the various morphological parameters, such as the amount and orientation of the crystalline phase and the orientation and state (rubber or glass) of the amorphous phase.

1 THE ROLE OF MOLECULAR STRUCTURE

This paper is not primarily concerned with molecular structure, but it is impossible to discuss "structure–property" relationships without beginning at this point. One of the main ideas behind this paper is that the molecular structure of a polymer does not uniquely define its solid-state properties and that the latter often depend much more strongly on the higher levels of structure ("morphology"). It must never be forgotten, however, that the morphology of a solid polymer, itself derives from the "interaction" of the molecular characteristics with the conditions under which the morphology is formed. We have only to remember that polymers as materials are what they are because they possess a characteristic molecular structure, namely that of a long chain.

†Lecture at the Scientific Symposium at the occasion of the Dedication of Midland Macromolecular Institute, September 29, 1972.

The proper perspective in structure–property relationships can best be given diagrammatically. In Figure 1 is shown a kind of "family tree" or line of descent, tracing the properties of polymers back to their molecular structure. To avoid over complicating the diagram, I have restricted it to the thermo-mechanical category of properties and omitted the electro-magnetic and chemical property categories. Clearly, however, these aspects could be included in a similar manner.

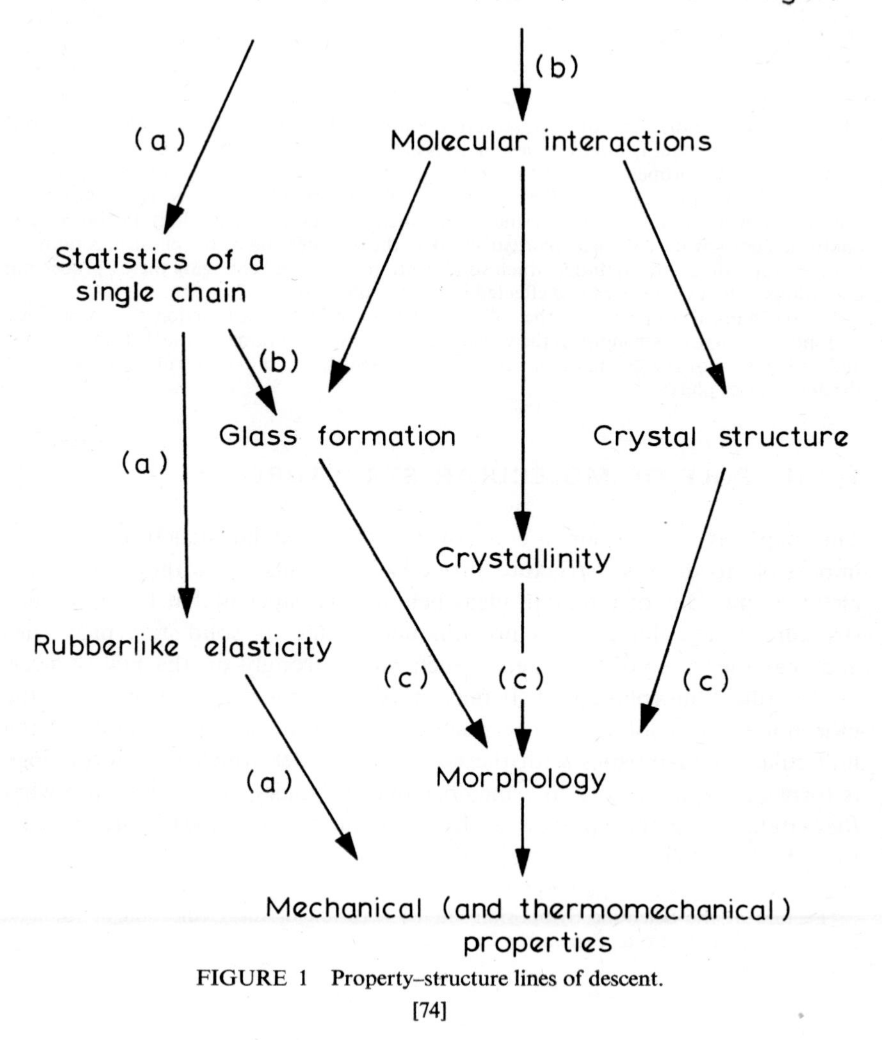

FIGURE 1 Property–structure lines of descent.

Reference to Figure 1 reveals immediately the very complicated nature of the dependence of mechanical properties (including thermomechanical properties) upon molecular structure. This is why many early attempts to relate properties directly to molecular structure met with only limited success. (My personal experience bears this out, since one of my first tasks as a newly qualified scientist in an industrial laboratory was to evaluate a series of aromatic esters for their fibre properties with a view to selecting the molecular structure which gave the best result. At that time very little was known about the physical structure or morphology of fibres.)

There is one property, that of rubberlike elasticity, which is largely dependent upon the independent behaviour of single molecules (route (a) in Figure 1). This behaviour is found at temperatures sufficiently high that the kinetic motion of the molecules swamps the constraint of molecular interactions. The material then behaves to a first approximation, as an assembly of non-interacting chains.

All other mechanical properties, however, depend strongly on molecular interaction. Indeed the temperature range over which rubberlike behaviour is observed is itself limited by the phenomenon of glass formation where molecular interactions predominate.

The role of interaction is indicated by routes (b) in Figure 1. According to the molecular structure, interactions result in the establishment of structures of a higher order, namely glasslike and crystal structures. Which of these two structures will form as a polymer is cooled from the melt or precipitated from solution depends both on the inherent capacity to crystallize and on the relative rates of cooling (or precipitation) and crystallization. Both capacity and kinetics are controlled by the molecular microstructure as well as by the more obvious thermodynamic parameters. Commonly both crystalline and amorphous regions occur in the solid polymer and their relative amounts and arrangements (size, shape, orientation, organization) depend not only upon the factors mentioned above but also on mechanical parameters such as stress or strain obtaining in the system during the formation of the morphology ((c) in Figure 1).

Finally, the mechanical properties are dependent directly on the morphology, since the response of the solid to stress or strain is the averaged sum of the responses of the individual components in the morphology. If the polymer is a homogeneous glass, the "morphology" must be construed in terms of appropriate concepts such as free-volume, close-range order and mean molecular orientation. If the polymer is a "composite" of crystalline and amorphous phases the situation is more complex, involving the totality of structural characteristics in both phases together with specific phase interactions (e.g. tie molecules anchored by the crystalline phase but extending through the amorphous phase).

It is clearly impossible to deal with the whole area defined by Figure 1 in a single paper. In what follows, the kinds of dependence discussed above will be illustrated mainly by examples drawn from our experience over some years of a single polymer—*cis*-polyisoprene or natural rubber.

Apart from the importance of this material in the history of polymer science, there are additional advantages of working with polyisoprene. It can be readily and quantitatively isomerized to change its microstructure; it crystallizes sufficiently slowly to allow control of the morphology; it can be obtained in a wholly amorphous condition for comparison with semi-crystalline specimens and, finally, the amorphous phase can be changed from glassy to rubberlike by passage through the glass transition temperature at the reasonably convenient temperature of −70°C, without changing the crystalline phase in any way.

The examples chosen below to illustrate structure property relationships are as follows. Firstly, the effect of small microstructural changes on the morphology and kinetics of crystallization. Secondly, the effect of strain on morphology, strain being regarded as a "processing variable", and thirdly, the effect of morphology upon some high-strain mechanical properties such as modulus, yield stress and fracture stress. The work is discussed here in review form only and further details will be found in the references. Acknowledgements are due to several co-workers, namely Drs. P. J. Owen, P. E. Reed, A. Singh and Ingrid Voigt-Martin.

2 THE EFFECT OF MOLECULAR MICROSTRUCTURE ON CRYSTALLINE MORPHOLOGY AND KINETICS

Final properties, we have asserted, depend on morphology and this, in turn, depends upon those kinetic processes which give rise to it. It follows that crystallization kinetics, interacting with the "processing conditions" (thermal history, for example) are of central importance to the subject. Andrews, Owen and Singh[1] have investigated the effect of small changes in molecular microstructure upon the crystallization kinetics of *cis*-polyisoprene. The changes were of two kinds, namely small amounts (up to 10 %) of isomerization of the initially *cis*-polymer and chemical cross-linking by peroxide decomposition involving equally small numbers of monomer units.

The linear growth rates of lamellar crystals (grown in very thin films of the melt) were measured directly using electron microscopy and an osmium tetroxide staining technique which arrests crystal growth after any chosen period of time. The dramatic results are shown in Figure 2. At a given temperature (−26°C) the linear growth rate, G, of the crystals is reduced one-thousand-fold by the random introduction of one isomerized unit in ten homopolymer units (10% isomerization). There is a linear dependence of log G upon the fraction β of isomerized material. A similar effect is produced by cross-linking although this appears to be only half as effective as isomerization in suppressing the crystal growth rate.

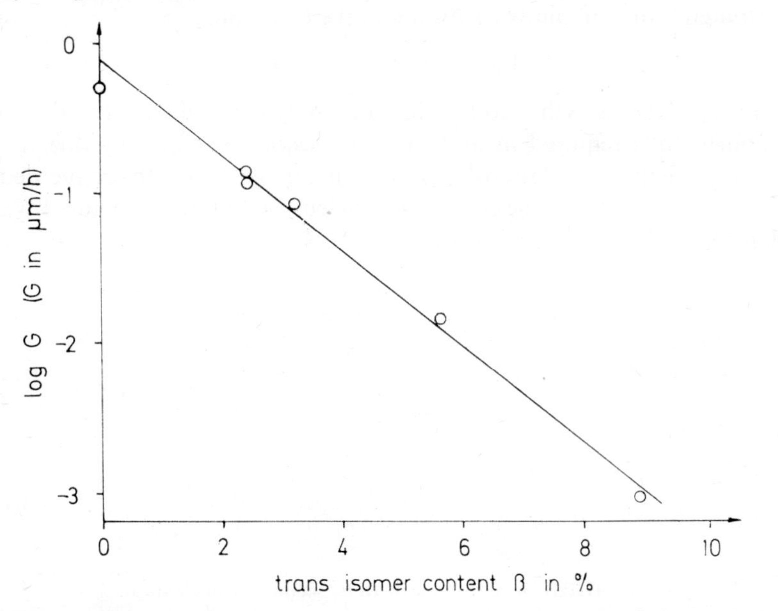

FIGURE 2 Single crystal growth rate G at $-26°C$, as a function of the fraction β of isomerized material in *cis*-polyisoprene [after Andrews, Owen and Singh;[1] used by permission of the Royal Society.]

Since the temperature dependence of G was unaffected by isomerization or cross-linking, it was concluded that the reduction in G could not be due to changes in the thermodynamic parameters (heat of fusion, melting temperature, interfacial energies) which appear in the classical equations[2] for growth by secondary nucleation. Instead, the authors explained their result in terms of a novel "exclusion" mechanism, as follows.

Continuous growth of a polymer crystal occurs by a modification of the classical "secondary nucleation" process. In this process, the slowest (and thus the rate controlling) step is the deposition of a new block of solid material on to a smooth growth surface (Figure 3). Once this block (the "secondary nucleus") has been deposited, a new monolayer is rapidly completed because atoms or molecules crystallize readily at any step on the surface. In polymer crystallization the process is probably more complex, involving the secondary nucleation of each separate molecule on to the growth surface. In either case, however, it is necessary to establish a secondary nucleus which is large enough to be thermodynamically stable. This "critical size" requirement suggested that an uninterrupted sequence of homopolymer units, sufficient in number to form a critical size nucleus, was a requirement for the attachment of a new molecule to the growth surface. Andrews *et al.* were able to show theoretically that the crystal growth rate G would be affected by the fraction β of randomly distributed "foreign" units according to the equation,

$$\ln(G/G_0) = -(N - 1)\beta \qquad (1)$$

where G_0 is the growth rate for the homopolymer and N is the number of monomer units required in uninterrupted sequence. Not only does Eq. (1) describe exactly the relationship found in Figure 2, but the value deduced for N gave a width for the secondary nucleus of 1.98 nm compared with an independent thermodynamic estimate of 1.94 nm.

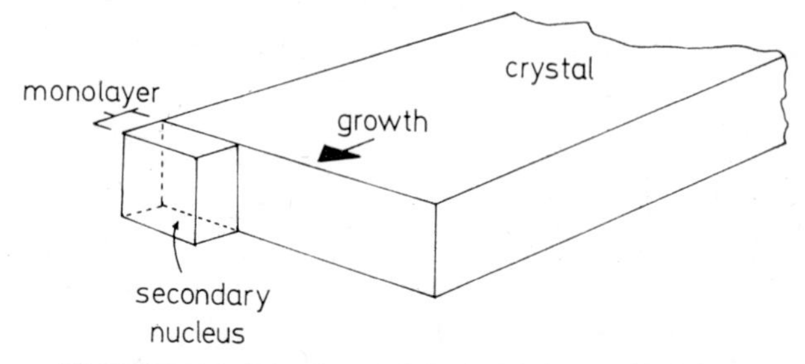

FIGURE 3 Crystal growth by secondary nucleation.

It appears, then, that the growth kinetics are extremely sensitive to molecular inperfections of whatever kind, the effects being far more marked than those of molecular weight variation,[3] for example. The amount and perfection of crystallinity achieved in a polymer subjected to a given processing history is thus likely to be equally sensitive unless care is taken to anneal the material.

The influence of small amounts of isomerization upon the ultimate (i.e. well annealed) morphology is negligible. Thus, provided time is given for the crystals to grow to their full extent, the results are wholly similar. In normal polymer conversion processes, however, such as extension, fibre spinning, blow moulding, etc., morphology may be strongly affected by changes in crystal growth kinetics. For example, regions which crystallize under high shear stress will have a very different morphology from regions which crystallize after the shear stress has relaxed. Rapidly crystallizing species may thus have a "high shear" morphology due to transient shear stresses whilst slowly crystallizing species may reveal little or no "memory" of the time spent by the melt under shear.

Cross-linked specimens of *cis*-polyisoprene, unlike isomerized material, *do* exhibit important morphological differences from the uncross-linked case.[4] These differences are illustrated in Figures 4 and 5. Figure 4a shows a spherulite grown in an uncross-linked thin film and Figure 4b one grown in a cross-linked film. In the latter, the lamellae are relatively short and discontinuous by comparison with the former. This may be due either to the physical impedance of growth by localized regions of high cross-linking or by an increased tendency for nucleation of lamellae resulting in a large population of "young" (and thus short) crystals. Figures 5a and 5b show a similar contrast for crystallization in a strained film (see next section). The long row-nucleated structures, often up to 10–20 μm, typical of uncross-linked rubber are strongly suppressed by cross-linking. Row nuclei still occur in cross-linked specimens, but are typically only a fraction of a micron in length, though much more numerous. The explanation of this may at first appear to be a physical blockage of row-nucleus growth, but is more likely to arise from the increased elasticity of the melt (which opposes the shape change involved in the formation of a long cylindrical nucleus).

3 THE EFFECT OF MECHANICAL CONSTRAINT ON MORPHOLOGY

3.1 Strain in the melt

During polymer conversion the melt is normally subject to deformation, either by shear in the die or mould, or by subsequent drawing or blowing processes. The degree to which these processes affect polymer morphology depends on the time, temperature and the severity of the deformation. If the stresses relax before crystallization is initiated, the latter occurs isotropically, but if stresses are present during even the early stages of solidification the effect on morphology can be profound.

Natural rubber films have a sufficiently high molecular weight to sustain applied stress in their melt condition even in the absence of cross-linking. They therefore afford a model thermoplastic system in which the effects of melt stress can be studied by the application of *static* strain (not possible in most polymer melts).

The results of this study are already well-documented,[5] and Figures 4a and 5a have already been given to show the morphological differences between a film crystallized unstrained and one crystallized from a strained (200%) melt. Kinetic evidence[1] reveals that the stress relaxes rapidly in the strained (uncross-linked) film, so that most of the crystallinity develops in a relaxed film. The *initial* crystallization, however, consisting of a row nucleus, forms rapidly under stress and governs the subsequent growth morphology even after the stresses have relaxed.

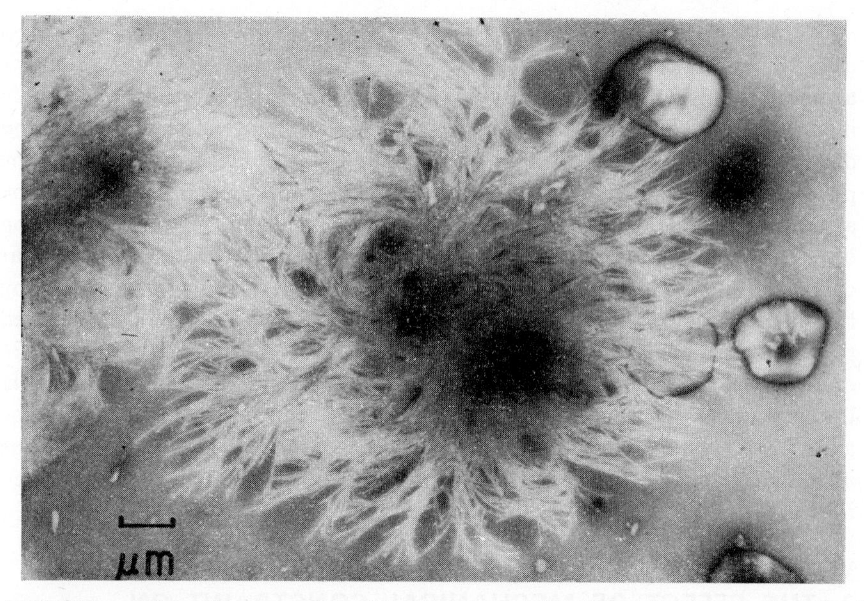

FIGURE 4a Spherulite grown in a thin film of natural rubber (all micrographs stained with OsO_4). [Reprinted by permission of the Royal Society.]

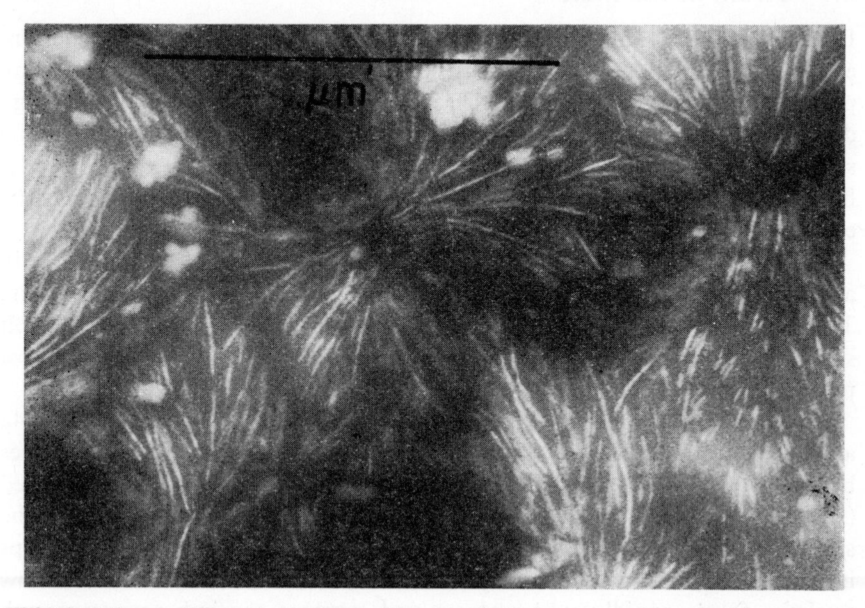

FIGURE 4b As Figure 4a, but in a film cross-linked by exposure to sulphur chloride vapour (after Owen[4]).

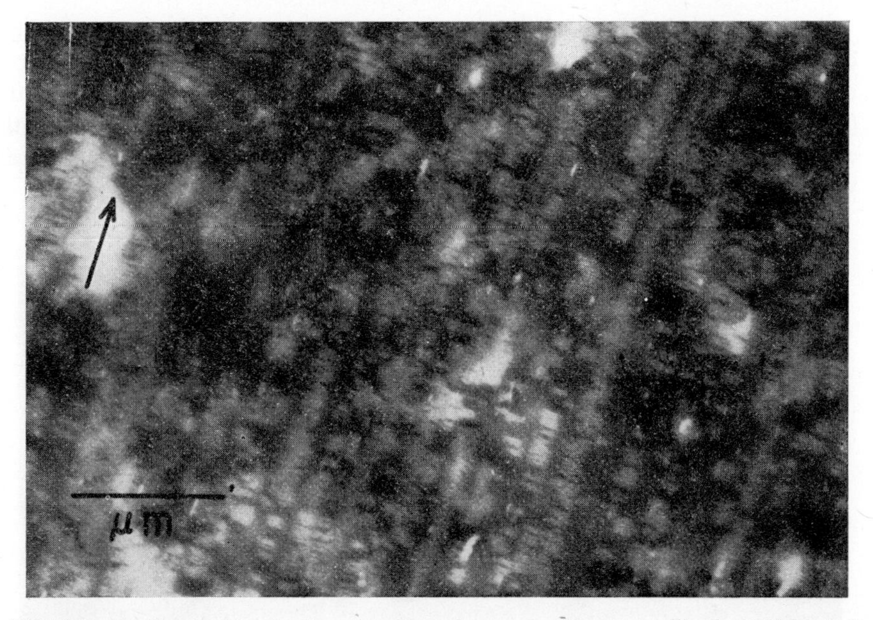

FIGURE 5a Row nucleation in a thin film of natural rubber crystallized at 300% strain (after Owen[4]).

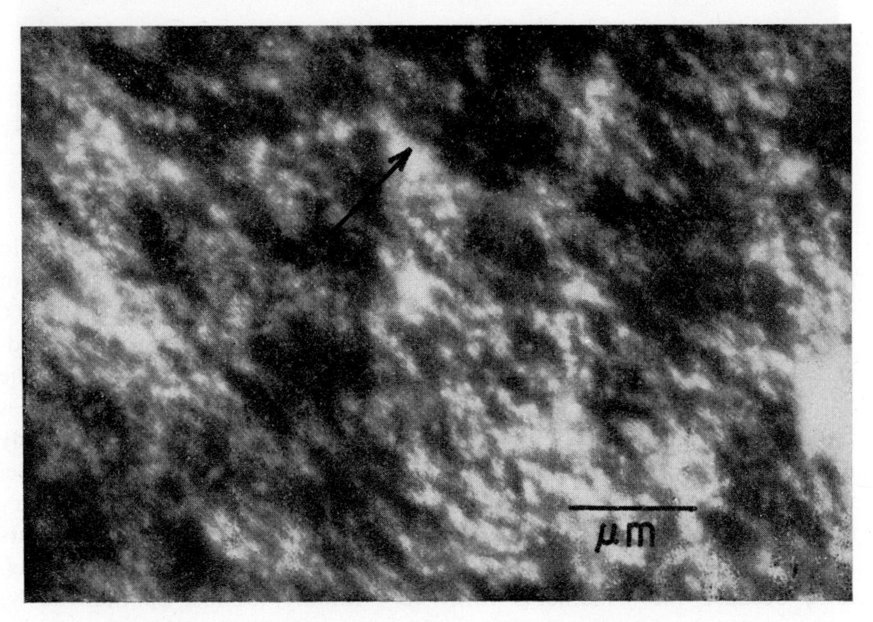

FIGURE 5b As Figure 5a, but in a cross-linked film (after Owen[4]).

At higher melt strains the density of row-nucleation increases, and the crystal melting temperature is elevated until spontaneous crystallization occurs at room temperature in a form described by Andrews[5] as γ-filaments (Figure 6). Alongside the obvious morphological changes associated with crystallization from a strained melt must be included the hidden morphological variable of molecular orientation in the amorphous phase.

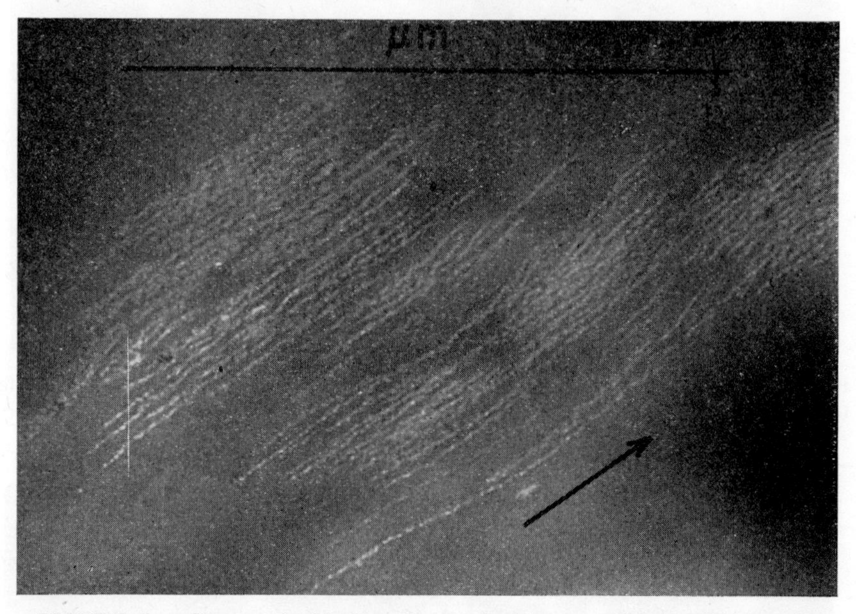

FIGURE 6 γ-filaments in thin film of natural rubber crystallized at 400% strain.

Figure 7 summarizes schematically the morphological effects of melt strain in natural rubber. In all cases the molecular chain axis is normal to the plane of lamellar crystals but lies parallel to the length of row nuclei and γ-filaments. This encourages the view that the lamellae are essentially similar to the folded-chain single crystals obtained by precipitation from dilute solution, and the variation in their thickness with crystallization temperature[2] reinforces this view. In the same way, row nuclei and γ-filaments are probably highly imperfect extended chain crystals as sketched in Figure 8. The row nuclei formed at intermediate strains (Figure 9 shows the early stages) are often very cleanly defined filaments of diameter as little as 3 nm. By contrast, γ-filaments are always particulate along their length[5,6] (Figure 6). This difference may well arise because the γ-filaments form very rapidly (almost spontaneously) whereas the rate of growth of row-nuclei at lower strains (although greater than that of lamellar crystals by a factor of 10^2) is still measurable[4] at some 100 μm/h. The degree of perfection is thus much lower in the γ-filaments.

The qualitative features outlined in Figure 7 are not restricted to *cis*-polyisoprene, but appear to be quite general for the melt crystallization of thermoplastics. Thus, almost identical patterns of behaviour have been found for isotactic PMMA, polycarbonate,[7] *trans*-polyisoprene,[8] *trans*-polychloroprene, isotactic polystyrene[9,10] and polyethylene.[9]

The quantitative aspects of row-nucleated morphology depend, of course, upon the amount of row nucleated material formed, since the row-nucleation density increases with time to a certain level and then remains constant. The final row nucleation density has been measured[4] for *cis*-polyisoprene as

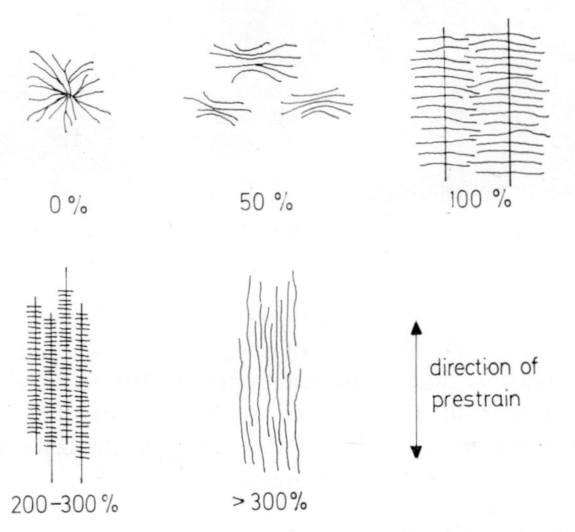

FIGURE 7 Morphological effects of melt strain in crystallized natural rubber (schematic).

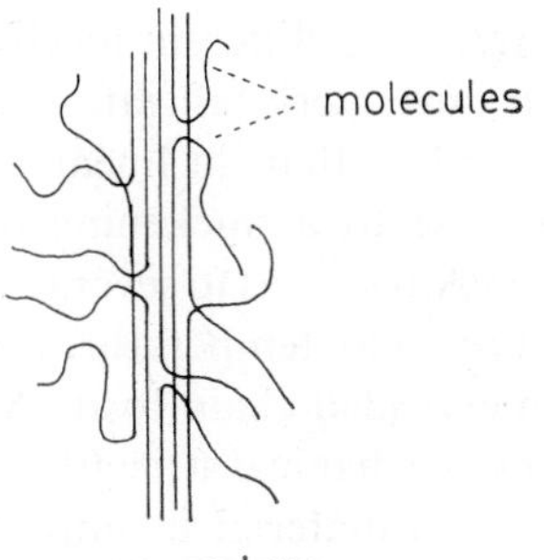

FIGURE 8 Possible structure of row nuclei and γ-filaments (schematic).

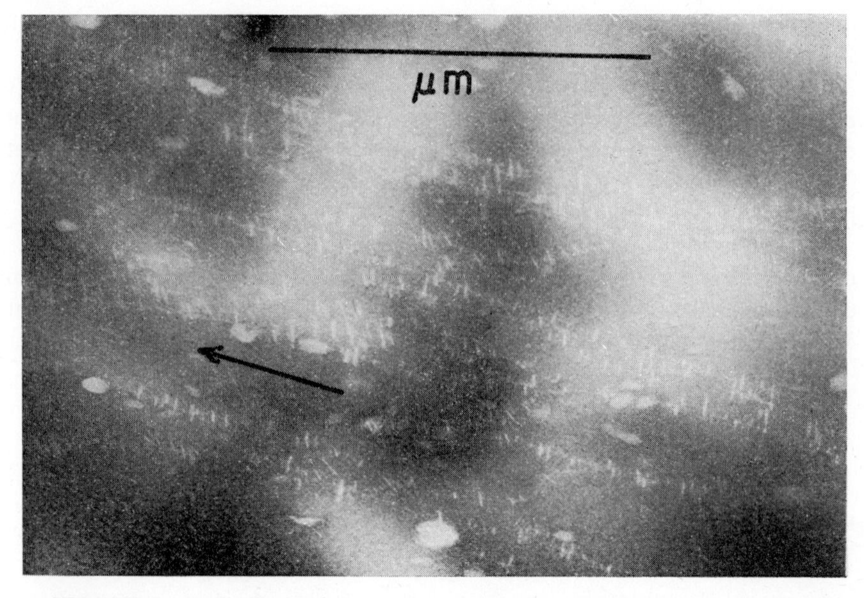

FIGURE 9 Early stages of row-nucleated crystallization showing row-nucleus.

a function of strain and temperature and is shown in Figure 10. The effect of strain is self-evident, but the effect of temperature is very interesting. As the melt becomes warmer the row nucleation density first decreases to a minimum around −12°C and then rises again. At high strains this rise is dramatic and clearly foreshadows the spontaneous γ-crystallization observed at slightly higher strains and temperatures.

3.2 Hydrostatic pressure

Finally in this section we refer to a different mechanical constraint which can be imposed upon the melt during crystallization, namely that of hydrostatic pressure. It has been established that, in linear polyethylene, crystallization under high pressure gives rise to a thickening of the normal folded-chain lamellae from a few hundred Angstroms to several microns.[11,12] The thickening at sufficiently high pressures and temperatures is such that the molecules unfold completely into an extended chain form. Material in this condition is extremely friable, i.e. has no mechanical strength, emphasizing the importance of inter-crystalline amorphous material in imparting the toughness usually associated with polyethylene.

Some preliminary experiments[13] have been carried out on the crystallization of thin films of natural rubber under gas pressures of up to 5 kbar. The most

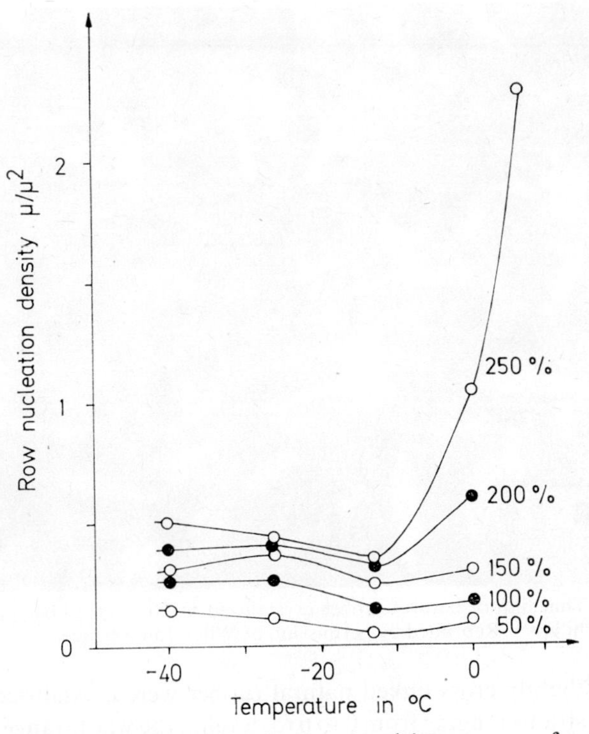

FIGURE 10 Final row nucleation density for *cis*-polyisoprene as a function of strain and temperature (after Owen[4]).

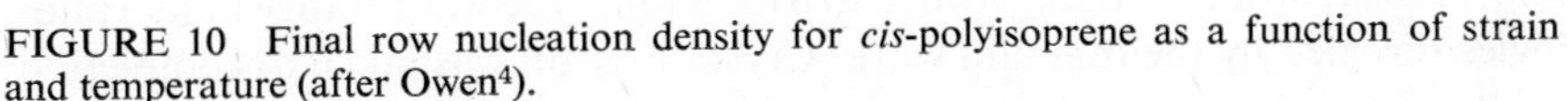

obvious effects are that the nucleation density is vastly increased so that unstrained films no longer exhibit a spherulitic habit but, instead, show a high concentration of separate single crystal zones shaped like oblate spheroids. Since there is good evidence[14,15] that the "aspect ratio" (length-to-thickness ratio) of lamellae in a bulk polymer strongly affect its elastic modulus, the destruction of the spherulitic habit by pressure could significantly affect mechanical properties. Needless to say, many polymer forming processes involve the application of hydrostatic pressure as well as shear stress.

4 MORPHOLOGY AND MECHANICAL PROPERTIES IN NATURAL RUBBER

4.1 The investigations

The investigations[16] detailed in this section, and carried out by Dr. P. E. Reed, were based on the thin film studies described in the previous section. Bulk

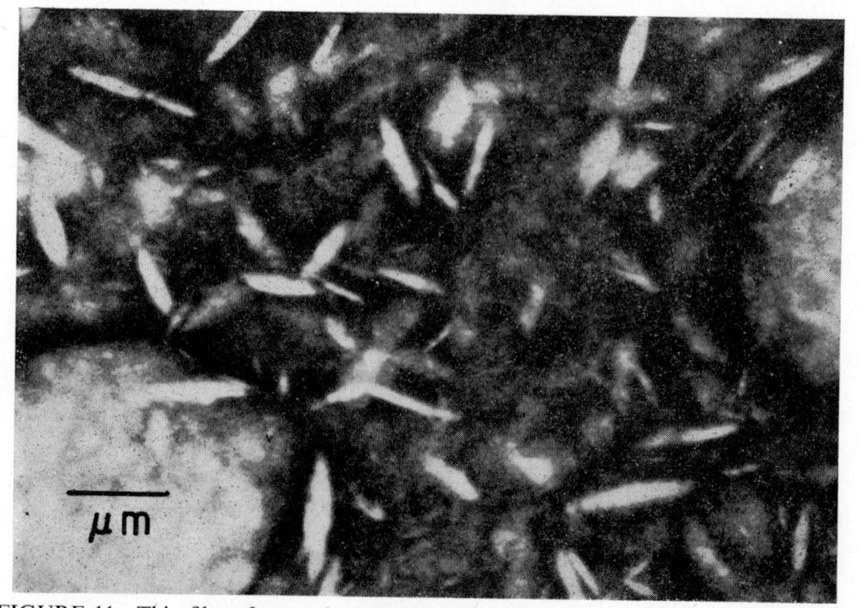

FIGURE 11 Thin film of natural rubber crystallized at 20°C at 4 kbar pressure. [After Andrews and Phillips.[13] Reprinted by permission of Wiley-Interscience.]

specimens of lightly cross-linked natural rubber were crystallized at −26°C under tensile strains ranging from 0 to 6 to develop the whole range of morphologies revealed by the thin film work. Crystallization was taken to completion in all cases giving a crystalline fraction of 0.3 ± 0.05 for all samples, and tensile tests were carried out at temperatures below −26°C covering both the rubber-like and glassy regions of the amorphous polymer (above and below −70°C respectively). From the stress–strain curves, mechanical parameters such as initial modulus, yield stress, fracture stress and fracture strain could be obtained.

Basic to this work is the assumption that the morphology in bulk is essentially similar to that in thin films for equivalent melt strains, apart from the "three-dimensional" character of the former. Thus, at zero melt strain we expect a spherulitic structure, at 50–100% spherulites flattened to bring their lamellar plane normals (molecular axis) towards the strain axis, at 100–400% cylindrical row-nucleated "shish-kebab" structures and at strains greater than 400%, γ-filaments. Because both lamellae and row nuclei are such small structures, thin sectioning of the bulk fails to reveal anything but a loss of spherulitic morphology with increasing strain. However, both light scattering[17] and X-ray[6] studies (of polychloroprene) provide strong indirect evidence that bulk morphologies correspond closely to their thin film counterparts at equivalent melt strain.

Reed's collected data for the mechanical properties of crystalline natural rubber is shown in a series of three-dimensional graphs, Figures 12 to 16. Each diagram shows one mechanical property (e.g. fracture stress) plotted vertically as a function of pre-strain (i.e. melt strain, the "morphological variable") and of the temperature of testing. The results will be discussed separately.

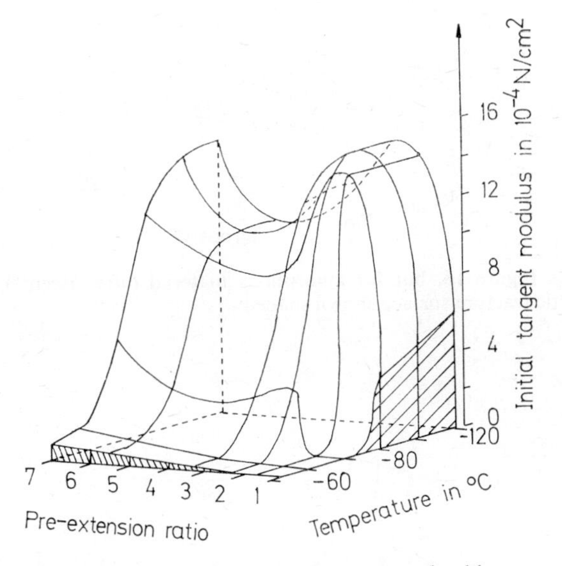

FIGURE 12 Initial tangent modulus of crystallized natural rubber as a function of pre-strain and temperature (after Reed[16]).

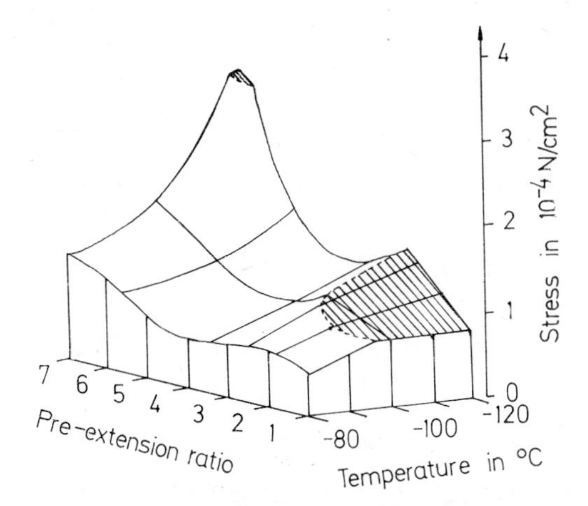

FIGURE 13 As Figure 12, but for yield and brittle fracture stresses of crystallized material (after Reed[16]). Yield surface shown plain, brittle fracture surface shaded.

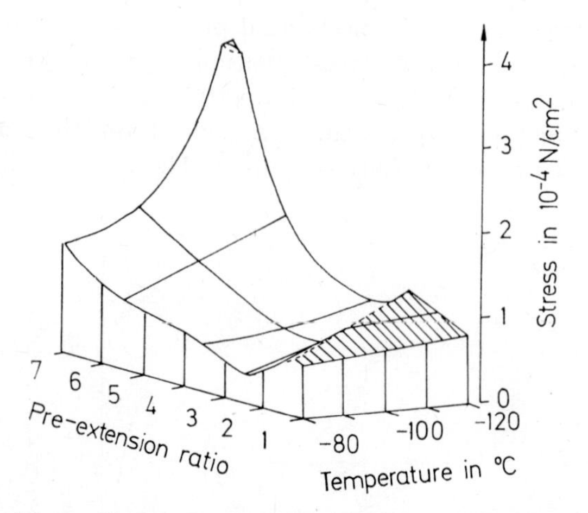

FIGURE 14 As Figure 13, but for amorphous material (after Reed[16]). Yield surface shown plain, brittle fracture surface shown shaded.

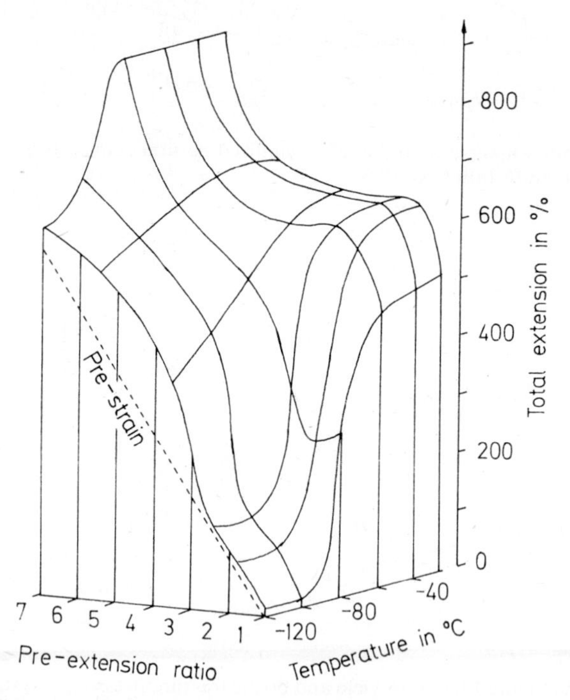

FIGURE 15 As Figure 12, but for total breaking strain (after Reed[16]).

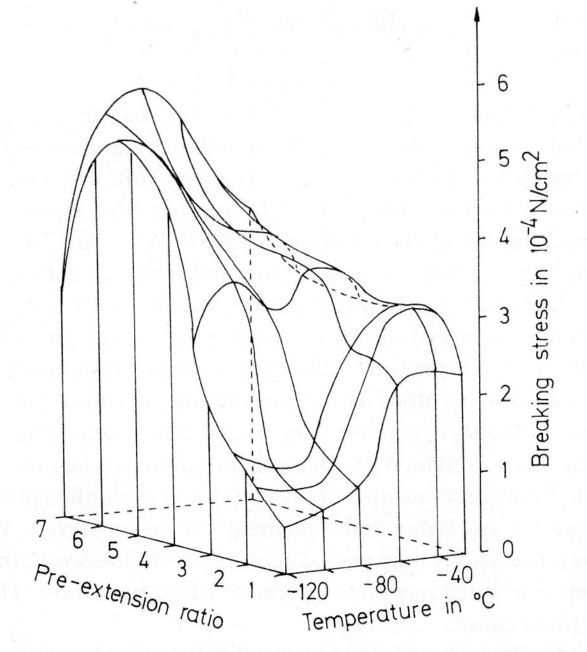

FIGURE 16 As Figure 12, but for true breaking stress (after Reed[16]).

4.2 The initial modulus

The initial tangent modulus to the stress–strain curve is shown for crystallized specimens in Figure 12. Spherulitic material (corresponding to pre-extension of 0% strain or extension ratio of unity) deforms in an essentially rubberlike manner above −60°C, although its modulus is an order of magnitude greater than that of non-crystalline material. (On the scale of Figure 12 both these moduli are of negligible magnitude). There is evidence from other sources[15] that the moduli of spherulitic polymers is strongly affected by detailed spherulite structure such as the lamellar aspect-ratio referred to earlier. The morphological evidence (Figure 4b) thus suggests that cross-linked rubber, with discontinuous lamellae, should have a significantly lower modulus in the spherulitic form than uncross-linked material. This accords with the observation that raw rubber increases its stiffness by at least two orders of magnitude when crystallized spherulitically compared with the tenfold change found here.

As the pre-extension increases, the morphology changes from spherulitic to row-nucleated and finally to γ-filaments, but the progressive rise in initial modulus above −40°C can be attributed almost wholly to the increased orientation of the amorphous phase.

At an intermediate temperature of, say −60°C, the effect of changing morphology is more dramatic, the modulus rising very sharply on the establishment of row-nucleated structures in place of spherulites, but thereafter remaining fairly constant. It is suggested[18] that this is due to a "fibre reinforcement" mechanism in which the "shish-kebabs" act like oriented stiff fibres embedded in soft matrix. The effect fails to manifest itself at higher temperatures because the amorphous matrix is too soft, and fails to transmit significant shear stress from one "fibre" to another (this stress transmission is important because the fibres are discontinuous). At −60°C the amorphous material is entering its glass transition range and losing its rubbery character.

Once the amorphous phase is fully glass-like (below −80°C), the dependence of modulus on morphology follows a pattern which is not greatly affected by further reduction in temperature. The glassy state modulus of spherulitic polymer increases only twofold as the morphology is transformed to a row-nucleated form. That this increase can be attributed wholly to amorphous orientation has been convincingly demonstrated by studies on polychloroprene.[6] Clearly the elastic constants of the glassy and crystalline phases are now sufficiently alike for no "fibre reinforcement" to be observed. What is observed, however, is a wholly unexpected *decrease* in modulus with the establishment of the row-nucleated morphology above 100% pre-strain. This could be due to any of three causes.

Firstly, Reed[16] has shown that crystallization, under constant strain, into a row-nucleated morphology results in a relaxation of molecular orientation in the amorphous regions between lamellae. The *amorphous* orientation thus actually decreases with increasing pre-strain during the establishment of row-nucleated morphologies. This would reverse the trend observed for small pre-orientations, but would not really explain the persistence of the downturn in modulus up to high pre-extensions.

Secondly, the lamellar orientation having now become uniformly transverse to the direction of testing, it could be argued that fibre reinforcement effects are minimized. Reed[16] found, however, that specimens tested at different angles to the direction of pre-extension showed very little difference in modulus at −120°C, even when the angle was 90°. This shows that no "fibre reinforcement" occurs with lamellae in a glassy matrix as already proposed above.

Thirdly, it is possible that the mutually aligned lamellae in row-nucleated structures are able to deform co-operatively by crystallographic mechanisms (e.g. [001] slip) to increase the compliance of the bulk. Until mutual alignment is achieved such deformations are relatively ineffective (e.g. in the randomness of the spherulite), but a stack of parallel crystals can deform "as one". Since the lamellae are reasonably extensive at intermediate pre-extensions they effectively separate the glassy matrix into "slices" perpendicular to the tensile axis, so that the matrix has relatively little ability to constrain the deforming crystals. As the row nuclei become more closely spaced with increasing pre-extension the horizontal continuity of the lamellar array diminishes and the matrix exerts a more effective constraint. This, together with the continuing rise in amorphous orientation, eventually causes the modulus to rise again at high pre-extensions.

This explanation of the dependence of modulus on morphology, although qualitative, is in reasonable accord with our present understanding of strengthening mechanisms in two-phase systems and opens the way to more quantitative investigations.

4.3 Yield and brittle fracture stresses

In the glassy region, below −80°C, the stress–strain curves of both crystallized and amorphous natural rubber exhibit either brittle fracture or yield followed by strain hardening, according to the temperature and pre-orientation. Reed's data[16] for crystallized and amorphous rubber respectively are given in Figures 13 and 14. For pre-extensions above 300%, where spontaneous crystallization occurs during pre-extension, the data for the amorphous glass have been extrapolated algebraically. The regions of morphology and test temperature over which brittle fracture occurred are shown shaded. The testing rate was held constant throughout.

The differences between Figures 13 and 14 are not great, indicating that brittle fracture, yield and strain hardening are all strongly affected by amorphous orientation in low crystallinity polymers. This is further borne out by studies[6] on polychloroprene at 18% crystallinity. At higher crystalline content, we have shown elsewhere[18] (for polyethylenes) that inter-lamellar attachments play a much stronger part in determining yield behaviour.

Closer inspection of Figures 13 and 14 naturally reveals differences between the crystalline and amorphous cases. At low pre-extensions the brittle-ductile transition temperature is suppressed by the presence of crystallinity from −80°C to about −95°C. This is almost certainly due to the contribution of some low-stress ductility by the crystals. The second difference is that the shallow minimum which occurs in the yield stress as pre-extension increases, falls at a pre-extension ratio of 4 for the crystallized material but at 2.5 to 3.5 for the oriented amorphous specimens. This may again be explicable in terms of amorphous orientation since, as pointed out earlier, the local orientation in the amorphous phase of row-nucleated crystalline structures relaxes to a condition appropriate to a lower-than-nominal pre-extension.

If λ_s is the pre-extension ratio, Reed[16] calculated that the residual amorphous orientation in a row-nucleated specimen after crystallization would be,

$$\lambda_{AR} = 4.41\lambda_s/(6.3\text{–}0.3\lambda_s)$$

which gives $\lambda_{AR} = 3.5$ for $\lambda_s = 4.0$ in good agreement with the shift of the minimum, at least at the lower temperatures. The occurrence of a minimum in yield stress thus appears to relate solely to amorphous orientation.

The third difference between amorphous and crystallized material is the rise in yield stress in the latter at large pre-extensions and low temperatures. This is perhaps directly attributable to the γ-filaments which we have suggested contain a large proportion of extended chain material. Whilst we have seen that crystals in a glassy matrix provide no reinforcement of modulus (because the elastic constants of the two phases are too similar), the γ-filaments *can* provide considerable strengthening when the glassy matrix begins to flow, providing of course, the γ-filament itself does not yield. This reinforcement-against-yield will be greater with γ-filaments than with lamellar structures because (a) [001] or chain-axis slip will be easier in the latter and (b) because gross tensile deformation causes buckling of lamellae and rapidly brings the [001] direction into the plane of maximum shear stress; in contrast, γ-filaments will not tend to re-orient as a consequence of matrix flow.

It is likely that the yield phenomena discussed here are all associated with shear yielding. At lower temperatures Reed and Natarajan[19] have shown that a new mechanism of yielding (that of crazing) comes into play in natural rubber, and the competitive nature of these two yielding mechanisms has been widely recognized in glassy polymers.[20]

4.4 Breaking stress and strain

Figures 15 and 16 show the breaking strain and stress respectively as functions of the usual variables for crystallized material. In the experiments, specimens are deformed in two stages, namely pre-extension to produce variable morphology and deformation during test. The variable of real significance is, of course, the total strain accommodated by the specimen before fracture and in Figure 15, therefore, is plotted this total strain given as the pre-extension plus the deformation during testing.

Plotted in this way it can be clearly seen that the fracture strain has a "plateau" value of some 600% over most of the experimental range except for two regions. The first of these is the region of brittle behaviour already noted at low temperatures and low pre-extensions. Here very little extension occurs during test, the total strain deriving almost wholly from the room-temperature pre-extension. The second exception occurs at high pre-extension and temperatures above T_g where total extensions up to 900% are consistently recorded. This improvement in extensibility occurs with a morphology of γ-filaments in a rubberlike matrix and almost certainly results from a fibre strengthening mechanism which inhibits crack propagation normal to the fibre (γ-filament) axis. This mechanism is familiar in materials such as wood and arises from the relative weakness of inter-fibrillar bond which encourages longitudinal fibrillation and blunts transverse cracks.[21] Reed[16] obtained striking fibrillation effects in specimens with very high pre-orientations.

Whilst the total extensibility of non-brittle specimens does not vary by a factor of more than 1.5 (and then only at high pre-extensions), the associated fracture stresses display drastic variations with morphological changes. This can be seen from Figure 16, where the load-bearing capacity nearly doubles at large pre-extensions as the temperature falls from −26°C to −120°C, and, at low temperatures, increases tenfold as the morphology changes from spherulitic to γ-filamentous. The breaking stress in Figure 16 is in all cases corrected for changes in specimen cross-section, i.e. it is the true load-bearing capacity of the material at the point of fracture.

At temperatures well above T_g, the breaking stress is fairly independent of pre-extension. Here the stress–strain curves are rubberlike and the morphology must undergo severe deformation before fracture. Since the actual fracture stress is independent of starting morphology, it is probably governed entirely by the amorphous matrix.

At low pre-extensions (up to $\lambda_s = 3$), the effect of reducing temperature through T_g is to produce embrittlement. Fracture here occurs in relatively poorly aligned material and the ultimate load-bearing capacity is understandably low. As pre-orientation increases, however, there is a rapid increase in the low temperature strength until (at $\lambda_s > 5$) reductions in temperature actually *increase* the strength of the material to a maximum before embrittlement once again causes a reduction. In the high strength region of Figure 16 the stress–strain curve, typically, exhibits yield and immediate, rapid strain hardening. As strain hardening becomes increasingly severe (with further reduction in temperature or increase in λ_s) the break in the stress–strain curve at yield is eliminated and brittleness again supervenes (Figure 17). This rapid strain hardening is typical of crystallized material and is not found in oriented amorphous polymers. It is thus attributable directly to the crystalline morphology.

The strength rise at −120°C with initial pre-orientation is apparently associated with a rise in yield stress (see Figure 17). Incipient yield appears to lead directly to fracture, and earlier these fractures were classified as brittle because there is no post-yield extension. Such a rise in yield stress can be attributed to amorphous orientation alone, although parallel tests on non-crystalline specimens showed that the latter always yielded at lower stress than the crystallized ones. Once yield is fully established, with significant post-yield deformation, the high breaking stress of the crystallized material is obviously a direct result of strain hardening (Figure 17) which occurs far less rapidly in amorphous material. What is the cause of this strain hardening?

It has already been suggested that yield in the crystalline phase is progressively suppressed as lamellar or "shish-kebab" structures give way to

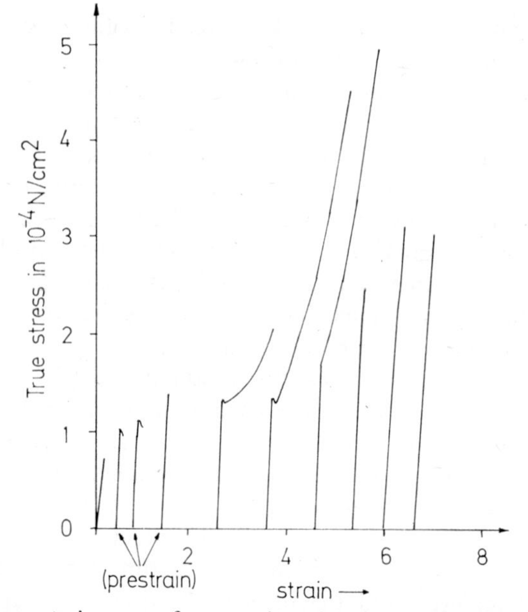

FIGURE 17 Stress–strain curves for pre-oriented and crystallized natural rubber tested at −120°C (after Reed[16]).

extended-chain or γ-filament fibrillar morphologies. Since the yield stresses for amorphous and crystalline specimens are not very different in the intermediate range of pre-extensions, it seems likely that here, at low temperatures, the onset of yield is controlled by the glassy phase. As significant plastic deformation accumulates, however, the glassy phase can only continue to deform if the attached crystalline material also deforms to accommodate the change of shape. Whilst this is relatively easy for lamellar crystals, it becomes progressively more difficult as the crystalline phase assumes a fibrillar character. This mechanism has already been invoked to explain the rise in yield stress at high pre-extensions and extends naturally to the post-yield region. A further factor may be the increased amount of inter-fibrillar connection as the γ-filaments become more closely packed with increasing pre-extension.

Eventually the strain-hardening mechanism is self-defeating; the material becomes so inextensible that it becomes prey to weakening by flaws, lacking any plastic flow to blunt incipient cracks. A new brittle condition is thus established and the strength falls drastically at $\lambda_s > 6$.

5 CONCLUSION

The detailed discussion of *cis*-polyisoprene has illustrated the validity of the original proposition, namely that the properties of polymeric solids depend upon their structure *at all levels* and not solely upon their molecular structure. The complexity of the data given in Figures 12–16 underlines the difficulty of providing simple rules in this area of structure–property relationships. Appeal has to be made to the anisotropic responses of both crystalline and amorphous phases to stress and, additionally, to specific mechanical interactions between the phases. It is in this latter area, for example, that molecular weight probably exercises a profound influence on mechanical behaviour.

The work reported here, then, represents just a fraction of the painstaking task of "mapping" properties in the multi-dimensional reference frame of structure. It is only as we continue this mapping that a rational and consistent picture will emerge.

One particular feature of the present paper is the necessity to appeal frequently to the concepts of composite material theory in seeking to explain the effects of morphology upon properties.

References

1. E. H. Andrews, P. J. Owen and A. Singh, *Proc. Roy. Soc. (London)* **A324,** 79 (1971).
2. J. D. Hoffman and J. J. Weeks, *J. Chem. Phys.* **37,** 1723 (1962).
3. J. M. Barrales-Rienda and J. M. G. Fatou, *Polymer* **13,** 407 (1972).
4. P. J. Owen, *Crystallization in Thin Films of Natural Rubber*, Ph.D. Thesis, Univ. of London (1970).
5. E. H. Andrews, *Proc. Roy. Soc. (London)* **A277,** 562 (1964).
6. E. H. Andrews and B. Reeve, *J. Mater. Sci.* **6,** 547 (1971).
7. J. J. Klement and P. H. Geil, *J. Macromol. Sci. (Phys.)* **B6,** 31 (1972).
8. C. K. L. Davies and Ong Eng Long, to be published (1973).
9. A. Keller and M. J. Machin, *J. Macromol. Sci. (Phys.)* **B1,** 41 (1967).
10. G. S. Y. Yeh and S. L. Lambert, *J. Appl. Phys.* **42,** 4614 (1971).
11. C. L. Gruner, B. Wunderlich and R. C. Bopp, *J. Polym. Sci.* A2, **7,** 2099 (1969).
12. D. C. Bassett and J. M. Phillips, *Polymer* **12,** 730 (1971).
13. E. H. Andrews and P. J. Phillips, *J. Polym. Sci.* **B10,** 321 (1972).
14. J. C. Halpin and J. L. Kardos, *J. Appl. Phys.* **43,** 2235 (1972).
15. P. J. Phillips and J. Patel, Private communications.
16. P. E. Reed, *The Influence of Crystalline Texture on the Tensile Properties of Natural Rubber*, Ph.D. Thesis, Univ. of London (1970).
17. W. Yau and R. S. Stein, *J. Polym. Sci.* A2, **6,** 1 (1968).
18. E. H. Andrews, *Pure & Appl. Chem.* **31,** 91 (1972).
19. R. Natarajan and P. E. Reed, *J. Polym. Sci.* A2, **10,** 585 (1972).
20. P. Bowden, in *The Physics of Glassy Plastics*, R. N. Haward, ed., Applied Science Publishers, U.K. (1973).
21. J. Cook and J. E. Gordon, *Proc. Roy. Soc. (London)* **A282,** 508 (1964)

Mechanical Fabrication of Thermoplastic Polymers†

T. ALFREY

The Dow Chemical Company, Midland, Mich. 48640

The physical properties and performance of fabricated thermoplastic items depend upon the molecular structure (mainly established during polymerization) and the spatial arrangement of the polymer molecules (established during fabrication). The spatial arrangements can be favorably influenced by controlled molecular orientation (uniaxial, biaxial and "crossed") and by composites (fiber reinforcement, multi-layer films).

The mechanical fabrication of thermoplastic polymers embraces a wide range of operations. Some processes, such as extrusion and injection-molding, are carried out at high temperatures where the polymers behave as viscoelastic fluids. Other processes, such as bottle blowing and vacuum forming, entail only moderate deformations, and can be carried out at temperatures where the polymers behave as viscoelastic solids. Some thermoplastic polymers are capable of being fabricated at still lower temperatures (below the glass temperature for amorphous polymers; below the melting point for crystalline polymers), by processes similar to the rolling, forging, and drawing of metals. In some processes, a crystallizable polymer enters the fabrication zone in a metastable, supercooled amorphous state, and crystallization occurs during the mechanical deformation.

The primary purpose of any fabrication operation is, of course, the establishment of a desired geometrical shape. A less apparent, but most important, feature which is simultaneously established is the spatial arrangement of the polymer molecules within the fabricated article. This includes orientation in amorphous polymers, morphology and orientation in crystalline polymers, and phase geometry in multiphase systems such as composites and foams

†Lecture at the Scientific Symposium at the occasion of the Dedication of the Midland Macromolecular Institute, September 29, 1972.

J

The physical properties and performance of fabricated thermoplastic items depend upon the molecular structure (mainly established during polymerization) and the spatial arrangement of the polymer molecules (established during fabrication).

Consider first the influence of molecular orientation in amorphous thermoplastics, using atactic polystyrene as a specific example. At room temperature, unoriented polystyrene is a brittle, glassy, amorphous polymer. Uniaxially oriented polystyrene is highly *anisotropic*. In the direction of orientation it can have a high tensile strength and ductile extensibility, and resistance to environmental stress-crazing and cracking.[1] On the other hand, in the transverse direction it is even weaker and more susceptible to stress-cracking than unoriented polystyrene.[2] The tensile strength in the orientation direction increases monotonically with increasing level of orientation, while the elongation-to-break rises to a maximum and then decreases.

Molecular orientation occurs inevitably in most mechanical fabrication operations, the pattern of orientation in the part being governed by the process *kinematics*—the pattern of flow and deformation followed by the polymer during fabrication. The *effects* of orientation can be either favorable or unfavorable, depending on the directions of orientation relative to the stresses encountered in service. Planned, controlled molecular orientation can be a valuable aid in achieving optimum properties and performance Uncontrolled orientation will have beneficial effects only by accident; usually it is a source of weakness and failure. Controlled molecular orientation in fabricated polymeric articles can be classified as uniaxial, biaxial, and "crossed".

Fibers and monofilaments are subjected in service mainly to tensile and bending loads; and consequently, uniaxial orientation can provide satisfactory mechanical performance. Natural fibers, such as cotton and silk, exhibit molecular orientation, and most synthetic fibers are subjected to controlled uniaxial orientation during manufacture.

Apart from fibers and monofilaments, most thermoplastic articles are subjected to multiaxial stresses in service. In films, sheets, and thin-walled vessels, biaxial stresses are usual. Consequently, uniaxial orientation is often detrimental in such items, whereas biaxial orientation can often be employed to advantage. Biaxially oriented polystyrene sheets (e.g. stretched 3:1 in both directions at 110°C) are strong and tough in all directions in the plane. Tensile strength, impact strength, and resistance to stress-cracking agents are all strongly enhanced, compared with unoriented polystyrene.[3] Biaxially oriented high impact polystyrene exhibits improved strength, impact strength, and fatigue resistance.

"Crossed" molecular orientation was employed by Cleereman, who rotated the core of the mold during injection-molding of thin-walled polystyrene tumblers.[4] The direction of molecular orientation varied through the wall, in a fashion similar to cross-laminated plywood. This led to a threefold increase in hoop strength, and a 10,000-fold increase in time-to-fail in a stress-cracking test in which tumblers were filled with corn oil or motor oil and pressurized. Such tumblers with crossed molecular orientation can be used as parisons for a subsequent bottle-blowing operation, yielding blown bottles with superior performance.

Molecular orientation is also important in crystalline polymers, although the pattern of behavior does not exactly parallel that exhibited by glassy amorphous polymers. For example, Kresser[5] has reported that the tensile strength of polypropylene systematically increases (approximately fourfold) with increasing uniaxial orientation, while the elongation-to-break systematically decreases. In any particular application, the optimum level of orientation represents a compromise in which both strength and toughness must be given consideration. Furthermore, in crystalline polymers, orientation can be introduced in two fundamentally different fashions:

1) A crystallized specimen can be stretched at a temperature which is higher than the glass transition point but lower than the crystalline melting point. This requires extensive reorganization of the original crystal morphology.

2) Molten polymer can be oriented, followed by crystallization of the oriented melt. The crystal morphology so obtained, and the resulting physical properties, are different from those obtained by stretching already crystalline polymers.

Clark and Garber[6] have emphasized the effects of mechanical processing on the morphology and properties of crystalline polymers, in the cases of blown film, injection-molded specimens, and spun fibers. For example, blown polyoxymethylene film exhibits row-nucleated ("shish-kebab") morphology, and much higher elongation than *spherulitic* POM. Injection-molding of POM yields a skin-core structure: a skin layer formed from multiple nuclei under high shear stress, during filling; a continued growth inward of twisting lamellae after mold filling; and a spherulitic core nucleated and crystallized under quiescent conditions.

The manufacture of Saran film involves a biaxial orientation step carried out in a bubble, at a temperature below the crystalline melting point. Molten (but crystallizable) polymer is extruded as a tube, and rapidly quenched past the crystallization range, This metastable, supercooled, amorphous tube is passed between nip-rolls into a bubble zone of trapped air. Here the tube undergoes both machine-direction and circumferential stretching, which induces crystallization. The product, a biaxially oriented, crystalline, thin-walled tube, is passed through converging rolls and nip-rolls, and either slit or wound up as a flattened tube. Machine direction orientation can be made greater, equal, or less than the circumferential orientation by controlling mechanical variables in the bubble zone. Mechanical properties in the two directions, and shrinkage characteristics, can be made balanced or unbalanced, as called for in various specific end uses.[7]

Such bubble processes belong to a class of operations which are susceptible to complete mechanical analysis—namely those involving membranes of rotational symmetry.[16] This class includes both batch operations such as vacuum forming and continuous operations. In all these operations, whether batch or continuous, certain common features are encountered—namely, the geometry of rotationally symmetric surfaces, and the Equations of Equilibrium for membranes of rotational symmetry. The meridianal direction and the "hoop" direction, by symmetry, are also principal directions of stress and strain (as well as of curvature). The third principal direction is normal to the surface; and the stress in this direction can be taken as zero. The stress at any point is thus specified by two components: σ_M and σ_H. These principal stresses are everywhere non-negative in the processes under consideration. Since we usually are dealing with large deformations of a nearly incompressible material, it is usually convenient to define the state of deformation in terms of the two principal extension-ratios, λ_M and λ_H. If the material properties are simple, a complete predictive mechanical analysis can be made, by combining the laws of geometry, the equations of equilibrium and the material properties. In the Saran film process, where crystallization accompanies deformation, such a predictive analysis is impossible. Even so, a complete mechanical analysis can be made by setting the process into steady state operation, and then making the following measurements: (a) the air pressure in the bubble; (b) the machine tension; (c) the meridianal curve, $r(z)$; (d) the location of a moving particle along the meridian, as a function of time.

With this information, the Equations of Equilibrium can be used to calculate $\sigma_M(z)$ and $\sigma_H(z)$. The extension ratios, λ_M and λ_H, can be calculated for every point in the stretching zones. The trajectories on the biaxial stress plane and the biaxial extension-ratio plane can now be plotted, with the aim of correlating them with the directional physical properties of the product.

Some thermoplastic polymers can be fabricated at relatively low temperatures, by methods similar to those used in metal fabrication (crystalline polymers below their melting points; amorphous polymers below their glass temperatures). These low-temperature operations, such as cold rolling,[8] drawing,[9] and forging,[10] invariably introduce large orientations which strongly influence the mechanical performance of fabricated parts. In low-temperature forming, the polymer must yield in a ductile manner, rather than fracture. This eliminates many thermoplastic materials; but sometimes mechanical pretreatment can convert an unusable thermoplastic to a usable condition.

For example, polystyrene is too brittle to cold-roll, but *oriented* polystyrene can be cold-rolled and further oriented.[8]

In multi-phase systems, such as composites and foams, another aspect of spatial arrangement is encountered. This is the geometrical disposition of the phases, which is established during fabrication.

The use of fibrous composites to achieve outstanding mechanical performance goes back to ancient times, and in recent years the role of fiber geometry has been the subject of intensive study.[12] In cross-plied laminates, the properties of a ply depend upon the mechanical properties of fiber and matrix material, and the geometrical arrangement within the ply; properties of the laminate depend upon the anisotropic properties of the individual plies and the geometry of their arrangement in the composite. The geometrical disposition of fibers in flexible composites, such as automobile tires, is as decisive a factor as in rigid structural composites.

Phase geometry is also important in laminar composites. Some swordmakers of early times hammered down alternate layers of hard and soft steel, obtaining blades which would take a fine cutting edge and yet were strong and tough.[13] Today, many multilayer thermoplastic film laminates are manufactured which exhibit combinations of properties which cannot be matched by any one of the constituent thermoplastic materials. For example, films containing alternating "hard" and "soft" layers can exhibit mutual interlayer mechanical reinforcement. Thin layers of a high modulus, low elongation material, sandwiched between thin adhering layers of a high elongation material, may be prevented from undergoing transverse fracture; and thereby forced to undergo large ductile deformations when the composite is stressed.[14] Such a layered composite has both a high modulus and a high elongation-to-break; it has a much higher work-to-break than any of the individual layers. It is possible to prepare multilayer plastic films containing up to hundreds of parallel layers, by co-extrusion from the melt. The geometrical arrangement of the phases in such a multilayer film is controlled by the flow-pattern developed during the mechanical fabrication operation. One method of developing such flow patterns has been described by Schrenk *et al.*[11] Different molten polymers are introduced as radial layers into an annular channel and pumped down the channel while the inner shaft, the outer cylindrical surface, or both are rotated. The molten polymer follows helical flowlines of varying pitch, and the initially thick radial layers are twisted into thin spiral laminae. The eventual layer distribution in the tubular film depends in a predictable manner upon the feed-port geometry, the volumetric rates of the various polymers, and the angular velocities of the two rotatable surfaces.

Finally, polymeric *foams* represent a broad class of multiphase systems whose properties depend upon phase geometry as well as upon the material properties of the bulk polymers. Expanded polymers range from elastomeric through semi-rigid (e.g., polyethylene) to rigid.[15] They can be high-density or low-density; open-celled or closed-celled; linear or cross-linked; amorphous or crystalline. They can be directly utilized as end-products, or incorporated into structures such as sandwich panels, which efficiently utilize their unique density-property characteristics.

Over the years, much progress has been made in the understanding of molecular structure of polymers, and in the controlled synthesis of desired molecular structures. Molecular weight distributions, degree of stereo-regularity, copolymer sequence distributions, and related molecular features are most important in determining mechanical properties. But in many cases the supramolecular structure, subsequently established by mechanical means, is equally important in determining the performance of a fabricated item.

References

1. K. J. Cleereman, H. J. Karam and J. L. Williams, *Mod. Plast.* **30,** 119 (May, 1953).
2. L. J. Broutman and F. J. McGarry, *J. Appl. Polymer Sci.* **9,** 609 (1965).
3. L. S. Thomas and K. J. Cleereman, *SPE Journal* **28** (4), 2; (6), 9 (1972).
4. K. L. Cleereman, *SPE Journal* **23,** 43 (Oct., 1967); **25,** 56 (Jan., 1969).
5. T. O. Kresser, *Polypropylene*, Reinhold, New York (1960).
6. E. S. Clark and C. A. Garber, *AIChE Materials Eng. and Sci. Div.*, Biennial Conference (1970).
7. W. R. R. Park, *Plastics Film Technology*, Reinhold, New York (1969).
8. L. J. Broutman and R. S. Patil, *Poly. Eng. and Sci.* **11,** 165 (1971).
9. H. L. Li, P. J. Koch, D. C. Prevorsek and H. J. Oswald, *J. Macromol. Sci.-Phys.* **B4** (3), 687 (1970).
10. K. F. Wissbrun, *Poly. Eng. and Sci.* **11,** 28 (1971).
11. W. J. Schrenk and T. Alfrey, Jr., *ACS Symposium on Coextruded Plastic Films, Fibers, and Composites*, Boston, April, 1972.
12. J. E. Ashton and J. C. Halpin, *Primer on Composite Materials*, Technomic Publishing Company, Stamford (1969).
13. G. Slayter, *Sci. Am.* **206** (1), 124 (1962).
14. W. J. Schrenk and T. Alfrey, Jr., *Poly. Eng. and Sci.* **9,** 393 (1969).
15. R. E. Skochdopole, Cellular materials, *Encyclop. Poly. Sci. and Tech.* **3,** 80 (1965).
16. T. Alfrey, *Appl. Polymer Symposia* **17,** 3 (1971).

Subject Index